SCIENCE
AND THE
THEORY OF GOD

Xavier L. Suarez

authorHOUSE®

AuthorHouse™
1663 Liberty Drive
Bloomington, IN 47403
www.authorhouse.com
Phone: 1 (800) 839-8640

© 2017 Xavier L. Suarez. All rights reserved.

No part of this book may be reproduced, stored in a retrieval system, or transmitted by any means without the written permission of the author.

Artist name: Danny Battle

Published by AuthorHouse 11/25/2017

ISBN: 978-1-5462-0484-8 (sc)
ISBN: 978-1-5462-0483-1 (e)

Print information available on the last page.

Any people depicted in stock imagery provided by Thinkstock are models, and such images are being used for illustrative purposes only.
Certain stock imagery © Thinkstock.

This book is printed on acid-free paper.

Because of the dynamic nature of the Internet, any web addresses or links contained in this book may have changed since publication and may no longer be valid. The views expressed in this work are solely those of the author and do not necessarily reflect the views of the publisher, and the publisher hereby disclaims any responsibility for them.

TABLE OF CONTENTS

Preface vii
Introduction xix
CHAPTER I: In The Beginning: Entropy 1
CHAPTER II: Of Angels, Animals And Men 17
CHAPTER III: Free Will And The Modern Social Sciences 57
CHAPTER IV: Fallen Man To Renassaince Man 75
CHAPTER V: Man: The Social Animal 92
CHAPTER VI: A Brief History Of Humankind 111
CHAPTER VII: Who Was The "Son Of Man" 149
CHAPTER VIII: The Kingdom Of Heaven 174

Author Biography 187

PREFACE

It is the first day of what Christians call "Holy Week," and I find myself mesmerized by what has been called "the greatest story ever told."

As part of a large and very well-organized religion, the Jesus narrative presents some attractive mythological morsels. No one can deny the story makes for great science-fiction.

But my method is primarily that of science alone, without the fiction. It uses reason, above all, to search for truth. It is the methodology of what used to be called "natural theology," in the mode of William Paley – of whom we shall say more later.

I suspend childhood beliefs and think rationally. I ask my reader to do the same, at least for the moment.

We will get back to faith-based ideas soon. Such ideas, and the insights they provide, are like the icing on the cake. They add gloss to the sketch I am trying to draw, which is based on provable facts, not faith.

If you remove yourself from your childhood beliefs, the entire narrative of what Christians call "Holy Week" is unbelievable. Think about it: In a span of five days there is a triumphal entrance by Jesus into Jerusalem, followed by an unexpected, horrific rejection of the man by his own people, leading to his torture and execution, after a sham trial.

The death is by crucifixion – the kind of capital punishment reserved for the most heinous offenders.

What an amazing story, I thought to myself, during Palm Sunday services. These folks are convinced that a mild-spoken country preacher started the week as a rock star and ended the week – for no compelling reason – as a pariah.

Even more amazing, the Sunday readings are a prelude to the really unbelievable part of the story: when the same rock-star-turned-pariah manages to resurrect and reappear as a ghost to his closest followers, adding elaborate, cryptic and even more grandiose touches to his prior stories and homilies.

Whatever the merits of the Jesus story, there is no denying that for the next two thousand years, that strange story has somehow been interwoven with the most enlightening of scientific discoveries, by the most enlightened of scientists and philosophers, into a cohesive framework that guides the most enlightened of modern societies.

My plan is to analyze the logic of that history as it shaped the minds of men, women and their societies. My analysis is buttressed by the most sophisticated, modern, and objective methods that science can muster.

Some will object that I don't have the credentials necessary to do this analysis. By way of resume, I can offer three degrees: in mechanical engineering, law and public policy – the last two from Harvard. Additionally, during the past decade I have written four books analyzing and ultimately finding coherence in the most diverse disciplines, including biology, genetics, astrophysics, anthropology, political science, philosophy, theology and economics.

Although none of the books have undergone a rigorous process of peer-review, they have all been reviewed by academics and professionals in the mentioned fields. Two of them – *On the Likely Origin of Species* and *The Wealth of a Nation* – were presented and critiqued at an academic forum hosted by the president of Florida International University (Mark Rosenberg) at his campus residence.

Others might argue that I cannot be objective on this topic because I am a practicing Catholic – which I suppose I am, though I prefer to say that I am "practicing to be a Catholic." Of course, I cannot rule out some amount of subjectivity. It is part of the human condition to be subjective.

So let me lay bare my own subjective background.

Growing Up in an Ultra-Catholic Home and Country

I was raised in a very Catholic environment, because my parents were both from very observant, orthodox families. In addition, we grew up in what can aptly be described as a Catholic island – a Caribbean equivalent of Ireland. There were other religious groups in Cuba, but the mainstream religious culture was Catholic.

Jews, Protestants, and Muslims had their own temples and houses of worship. They followed their own calendars. But the prevailing calendar was the Catholic one, including Easter, Christmas, Good Friday and the forty days of Lent, when we were expected to do some kind of self-denial, to remind us of the sufferings of Jesus.

Of all the Christian denominations, the Catholic way of life seems the most convoluted, with its prescriptions for distinguishing "mortal" and "venial" sins and observance of fasting during Lent and abstaining from certain foods at certain times – e.g., from meat on Fridays.

Fast and abstinence rules were already being diluted during my childhood. But that was not the case for mortal sins – which had to be confessed to a priest in order to avoid eternal damnation.

Failing to obey one of the commandments was decidedly a mortal sin. And for most youngsters, the commandment that barred our way to salvation was the Sixth Commandment, the one that prohibits "adultery." Except that the translation we used said "thou shall not fornicate." That meant no sex except in marriage.

To this day, in movies and books, the best explanation a girl can give for not having casual sex is that she was raised as a Catholic. Guys understand. Sex outside of marriage is considered a mortal sin by most Catholics of the strict-observance kind.

Luckily for Catholics, there is a prescription for forgiveness that is both spiritually and psychologically rewarding. It is called "confession" or "reconciliation." When practiced at too young an age, it can have humorous results.

A Youngster at Confession

A picturesque story of the Catholic sacrament of confession was told to me recently by an aide. For some reason that I don't remember, I was explaining that the ritual of confession had value even as a secular institution. I quoted from Lee Iacocca's autobiography (titled "Iacocca") in which he argues that the mere fact of baring your soul to another human being is a healthy exercise in critical introspection.

(Which it is, as any clinical psychologist can attest.)

But back to my young aide. This evidently impressionable young lady regaled us all with the story of when she was in a parish school, perhaps all of 8 years old. She had been passing the food line at the school cafeteria when she saw what is referred to as a Caesar's salad, with the usual lettuce and tomatoes, covered with cheese and croutons.

The temptation to grab a crouton took hold of her, and she divested one of the salads of a single crouton, which she quickly devoured. But a nun caught her *in flagrante delicto* and scolded her.

At the first opportunity to confess, she blurted out to the priest that she had "stolen a crouton," no doubt thinking it was a violation of the commandment that says "thou shalt not steal." The amused priest explained that the scolding was not for violating a commandment or stealing the crouton, but because it is unhygienic to lay your hands on food that someone else will later eat.

For observant Catholics, the guilt accompanying sex outside of marriage or even having an abortion can be cured by confession. But how do they deal with the ongoing guilt of being divorced and remarried? That is a topic being addressed by the current pope (Francis), who is doing his best to reform what is seen as a rather iron-clad prohibition of divorce in Catholic dogma.

And that brings me to the question of what is Catholic dogma, and the related question of what is the essence of Christian belief, buttressed and premised as it is by Jewish beliefs.

This book is in great part devoted to that question, in an effort to develop a plausible, holistic understanding of what the creator probably intended when she/he/it sent the galaxies flying and burning in such a way that the Planet Earth would someday be molded into a habitable venue for our species.

The Judeo-Christian Belief System

The Judeo-Christian belief system is unique in its ability to answer the three big existential questions that all humans face; let me explain what those questions are.

One: Are Humans Flawed? It is hard to disagree with the notion that humans have some sort of built-in defect, something that should not have been there from the beginning. Jews, Christians and Muslims have a simple explanation for that: They believe that humans are flawed because of some kind of transgression committed by our ancestors. That transgression or "sin" contaminated our species and is passed on to our kids. It requires all kinds of motivational constraints, such as punitive laws for transgressors, taxes for income-producing adult citizens and norms of discipline for children that require minimal schooling, immunization from disease, etc.

Two: Who or What Created the Universe? Jews and Christians are convinced that the same almighty being that caused the universe to begin (the same primal force that philosophers call the "First Cause") propelled the stars and planets at the beginning of time. Jews, Christians and Muslims (plus just about every other world religion) also profess the belief that the appearance of our species required a second creative event. In other words, most religions preach that humans did not totally "evolve" spontaneously from the big apes; at some point, the First Cause intervened to add a unique "mind-psyche" that enables men and women to make choices, feel remorse, and even give up their lives for a stranger.

Three: There Is a Solution to the Disconnect Between Humans and God.

Through obedience to a God, humans have the ability to overcome the tendency to do bad things and ultimately get along with others well enough to conquer disease, eliminate hunger and avoid war. Some Jews and all Christians and Muslims believe that the "re-connection" ("religion") between our species and the almighty creator allows us to eventually share a joyful eternity with our maker.

The Judeo-Christian-Muslim construct, insofar as it is able to answer the three existential questions above, is not just unique. It is also uniquely compatible with science and reason. That is my unequivocal conclusion after studying these things for half a century.

I immersed myself in that study because I had made a decision that these questions are too important to just accept what my parents taught me.

As the reader will see, I have broken with my parents' thinking on a lot of issues; including fundamental tenets of political science, psychology and theology. But it is interesting that on the basic principles of Judeo-Christianity, the science I have learned is quite consistent with that belief system.

And yet I don't consider my parents, or their parents, to be my intellectual mentors. The closest thing to an intellectual mentor for me was my godfather, Ignacio Warner. Before his death, and for about two decades, we met once a week in his Key Biscayne home. He lent me or bought me books on every topic from anthropology to theoretical physics.

"*Tio* Warner," as we called him, was the most interesting man I ever met.

We usually got together on Tuesday evenings in glorious Key Biscayne.

Tuesdays with Ignacio

My godfather was a very well-read man, with an extraordinary memory and a grasp of both philosophy and science that was quite unusual, given the modern tendency towards specialization.

He was conversant with most fields of science, including physics, chemistry, thermodynamics and advanced concepts of structural and chemical engineering. In that sense, he was no different from any engineering graduate of MIT or Villanova (my own *alma mater*). But his knowledge went way beyond the empirical sciences and advanced engineering, which he had studied at Louisiana State University.

Besides his vast technological, post-graduate training, my godfather was also quite a historian, and an accomplished theologian and philosopher. He also knew a lot about arts, music and the rest of the humanities. He was what we nowadays call a "Renaissance man."

He was also anti-clerical. He did not like the ecclesiastical trappings of power; he particularly disliked Papal vestments and Papal titles, like "supreme pontiff," which he knew were borrowed from Roman military lore. He would have preferred the simplicity of Pope Francis over what he considered the pomposity of his predecessors. He was particularly critical of any Catholic dogma that he reasonably concluded was extraneous to the basic teaching of Jesus.

To a great extent, I shared that skepticism. Premised on that commonality of views, we decided to discuss all important matters on the basis of the simplest tenets of the biblical account, including and most notably Moses's Ten Commandments and Jesus's Sermon on the Mount.

We did that over rum and whiskey on Tuesday night sessions that I called "TOE's" (for "theory of everything"). These were no-holds-barred sessions in which we discussed science, history, government and religion.

My godfather was quite skeptical of church hierarchy and church dogmas. Artificial obligations like the Sunday mass were anathema to him. Yet he felt that he ought to go to mass at least once a week, and partake of the ritual, the history, and the richness of a tradition that went back, uninterrupted, for 2,000 years. Therefore, he went to church on Saturday at noon, which did not satisfy the rules of Sunday observance.

Although he was skeptical of religious institutions, per se, *Tio* Warner understood the sociological merits of Judeo-Christianity. As a connoisseur of art history, philosophy and science, he knew full well that Judeo-Christian values had been the bulwark of civilized behavior and scientific discovery, without which there might never have been a Renaissance. (I will make that argument more fully in Chapter VI.)

And so it was that we set about extracting the chaff from the wheat, without paying much attention to the chaff. We were looking for the kernel of truth in religion and science. Being engineers, we sensed that the universe was a marvelous construct, a fine-tuned, biological, organic, dynamic machine that by all rights should produce happiness for its inhabitants.

In that vein, my godfather and I assumed as little as possible of the conventional wisdom. We agreed that there was a God, an infinite being – a "Big Banger" if you will. *We agreed that the Big Banger decided (for reasons we could not fully grasp) to create a being with "free will" who did not have to follow the master plan.*

We also agreed that there was a good likelihood that the creature with free will at one point made one (or more) bad choices. And that the result of those bad choices was some sort of mutation in the human genetic tree.

A fault was found. Humans, henceforth, would have a hard time communicating, a hard time getting along, a hard time giving birth, a hard time sharing the resources of this earthly planet. Society was never going to be the same.

That was all we assumed – not because it was in the bible, but because it seemed self-evident to us.

The rest was science.

And we both reveled in late-into-the-evening discussions of science, history, philosophy. We wanted to know how everything started and later seemed to go awry.

We also sensed that things took a turn for the better when certain Middle East prophets proclaimed the need for a Messiah which we understood was a spiritual leader. That spiritual leader, based on the best historical evidence, had propelled a marvelous reform that reversed what had, up to then, seemed an irreversible course towards a world based on "might makes right" instead of "right makes might."

Before I start on the actual story, I need to explain clearly my methodology.

Aquinas: The Argument from Authority Is the Weakest

As I said, *Tio* Warner instilled into me a measure of skepticism regarding any "doctrine" or "dogma" that does not adhere to what is provable by science. I think the reader will appreciate that it has helped my analysis enormously. I feel competent to point out when there are apparent contradictions between the discoveries of modern science and the Judeo-Christian belief system.

Without knowing it, *Tio* Warner and I had both concluded, with Thomas Aquinas, that "the argument from authority is the weakest." This is a very important principle, and I need to dwell on it for a few moments.

Catholics (as well as the other of the more rigid offshoots of the main religions) too often base their views on what they consider a "higher authority," such as a pope or patriarch or supreme ayatollah. If they accept any notion of divine inspiration, or claim to infallibility, the dogmatic effect is even greater.

Whatever the merits of that approach, it is not the one I use. Like Aquinas, I recognize the convincing value of reason, logic and the scientific method. I couple those methods of ascertaining truth with what is perhaps the most practical and definitive one, particularly in the social sciences: the shared, learned experience provided by history.

It is very important, at this point, to acknowledge that "experts" are not necessarily the elite judges of what are the most fundamental truths. Experts have failed us in many moments of history. (Later in this book I will touch upon the macabre philosophy of one ostensibly great thinker, whose views justified Nazism and eugenics.) In discovering truth, one can never ignore the wisdom of ordinary people; even less can we ignore the accumulated wisdom of humanity's repeated efforts to achieve individual and collective welfare.

Here's how biologist Douglas Axe explains that common-sense wisdom:

> *Basic science is an integral part of how we live. We all make mental notes of what we observe. We all use those notes to build conceptual models of how things work. And we all refine these conceptual models as needed. Without doubt, this is science.*

Nothing – absolutely nothing – that fails the test of reason or history will be allowed to cloud our judgment as to what constitutes the truth. At the same time, no insight that comes from "scripture" or other allegedly prophetic pronouncement will be disregarded. In particular, the proven, benevolent history of two of the world's religions (Judaism and Christianity), as interpreted by the best thinkers and philosophers, will be holistically combined with secular science.

That is my methodology. And, I should add, it meets the test enunciated by physicist and Nobel Laureate, Percy Bridgman, when he said that "the scientific method, as far as it is a method, is nothing more than doing one's damnedest, with one's mind, no holds barred."

As for those who are believers, you will be comforted by my earnest, "no holds barred" conclusion that there are not many contradictions between modern science and modern theology. These two domains of human thought are not only consistent, but quite complementary. Each supports the other enormously – like the gears of an engine that mesh.

And so the reader will hopefully conclude, with this humble author, that history is similar to a vehicle that traveled from a standing start (the law of the jungle in primitive peoples) to the high gear of modern technology, democracy, and nearly universal freedom and welfare. But now I am getting ahead of myself

At no time will I expect those who read this to suspend their instinctive disbelief of things we cannot see and touch and measure. At no time will I attempt to impose any bias or prejudice to prove my point.

And I certainly hope not to bore my reader with tedious, technical explanations or complicated equations. Albert Einstein counseled us to "make things as simple as possible, but no simpler." I will do my best to follow that counsel.

The universe is complex, but not overly complicated. Perhaps for our benefit, it was assembled beginning with a little spot (or "singularity") on which a bunch of matter was placed and later, in a fraction of a second, organized into a tiny erector set, complete with rubber bands that snapped when released.

For almost fourteen billion years, we have been experiencing the expansion of that tiny mini-universe. Luckily, we can explore it rather well, from almost the first moment of its existence.

We can also examine the most sophisticated of the creatures that inhabit the earth. That creature (humankind) is relatively young, by cosmic standards. And it is one that contains a lot of information that is accessible to us by simply meditating. (That sort of reasoning process is called "deductive" in order to distinguish it from analysis done through measurements, which is called "inductive.")

We humans are all rather alike. We have slightly different heights and weights; roughly half of us are male and half female. Our skin color and hair texture vary slightly, but our blood and nervous systems are nearly identical. Our DNA's are essentially the same, with one extra chromosome depending on sex.

As we look at our bodies and reflect with our minds, we can come to conclusions about our nature, our longings, our yearnings to live happily.

Two modern sciences have fashioned tools with which to study human behavior. Psychology studies the individual and sociology studies the collection of individuals interacting. Those sciences did not exist in the Middle Ages or even during the Renaissance. They were formulated during the last couple of centuries, coinciding with a period of enormous advance in the natural sciences.

Neither Aristotle nor Aquinas in the thirteenth century, had any inkling of the discoveries in genetics, microbiology, atomic physics, or cosmology. None of them had at their disposal modern, statistical studies that map the way in which our minds interact with our bodies in what are called "psychosomatic" effects. None could have guessed that some of us are born with a defect that makes us prone to have emotional highs and lows, called "bipolar episodes." Or that some are born with a propensity to fold into our interior shells and have difficulty interacting with others (a condition referred to as "autism").

If the ancients could resurrect and have a forum in which their brilliance could share insights with modern scientists, it might take the shape of something that I describe in the section that follows.

Any reader who doesn't care to follow me into a whimsical, theatrical, fictional gathering of academics and savants covering 3,500 years of human history can skip the introduction and move to the first chapter. Besides some entertainment, it is also a sort of bibliography of authors on whom I have relied in crafting my own scientific analysis of what has been called the "God Theory."

For if there was always a God, possessed of the intelligence and the benevolence we now attribute to her or him, it makes sense to think that there was a good reason for creating not just galaxies, but human beings. And if there was a reason for creating human beings, it stands to reason that there was a benevolent purpose, rather than just a fanciful, instinctive, almost sadistic intent to frustrate that new creature with limitations, disease, war, and the angst of never experiencing absolute joy.

The mind of God, if there is a God, must be more explainable, more reasonable, more visionary than to create a flawed being.

And so we delve into the analysis, borrowing from as many of the great thinkers as possible to fit into one little book, written by one amateur "natural theologian," who uses the methodology of science exclusively as the ultimate yardstick for finding truth.

INTRODUCTION

One of the fascinating things about Augustine is that while he began his project of understanding God using human reason as his exemplar, he ended up understanding human reason by using God as an example instead.
Dr. Ferdinand Santos

The setting was Harvard University, and the intellectual forum took place at the beginning of the twentieth century. Two semicircular tables were set up - facing each other. At each sat three people.

On the far side sat Aristotle, Augustine and Aquinas. Opposing them were Sigmund Freud, William Paley and Pierre Teilhard de Jardin.

The idea of the forum was that the three great philosophers had asked to interview three modern thinkers on the interface between psychology, philosophy and natural theology – which, as previously defined, is a branch of science that explores the way in which empirical data, and theoretical equations that flow from the data, seem to suggest the existence of a creator or "Big Banger."

They had agreed to constrain the philosophical discussion to the commonly accepted tenets of classical philosophy, and thus to leave out existentialism, nihilism, Marxism and the strange mélange resulting when Darwinian principles were extrapolated from biology and fused into the pronouncements of Nietzsche to create "Social Darwinism," with its Hitler-macabre implications of racism and sexism.

This was a discussion among reasonable people, after all. The participants were expected to agree on classical philosophy and were hoping to put out an up-to-date position paper on how modern psychology and modern empirical science modified and enlightened the traditional theology embraced by most civilizations. That sort of scientific enlightenment of theology had been championed by the British philosophers of the nineteenth century, led by William Paley.

Aristotle, Augustine and Aquinas felt that modern methodologies should be applied to their own now-ancient ideas, which were solid enough, but needed touching up. They took it upon themselves to invite a modern psychologist (Freud), a natural theologian (Paley) and a thinker (Teilhard de Chardin) who had done his best to bring science and religion together – albeit with mystical overtones that resembled the Eastern philosophers more than the Western ones, of which they considered themselves the best three of all times.

Modesty was looked at differently by these great thinkers. In fact, there was a bit of a logistical problem from the start, as Aristotle felt that Aquinas should preside over the forum, whereas Aquinas felt that Aristotle should preside. Neither wanted to occupy the middle chair, in deference to what each considered the superior rank of the other.

Augustine, who had lived among saints and sinners like perhaps no one else in recorded history, possessed gobs of earthiness and common sense. Listening to their deferential argument (and somewhat humbled that neither of them had suggested him for the middle seat), he solved the dilemma as Solomon might have done. "Gentlemen," he said, "we should not quibble about rank; I will sit in the middle, and announce that the sitting arrangement simply reflects chronological sequence."

On the other side of the room, the sitting arrangement never caused any concern or dispute. Freud's *ego* prompted him to grab the middle chair; and he assumed that he deserved it, being much better known than the obscure Briton and the mystical Frenchman with the complicated name.

I should add that the participants had wrestled with the idea of making it a panel of five, rather than three. Aquinas had suggested the addition of Maimonides and Averroes. But they had a hard time contacting Averroes (whose real name was Abū l-Walīd Muḥammad bin 'Aḥmad bin Rušd). And Maimonides had begged off, saying that he was a strict observant of the Sabbath, and that this being a weekend symposium would prevent his full and continued participation. "Anyhow," he had added, "I don't have any major discrepancies with your theories and will be available with my friend Averroes, via teleconference – if needed."

Thus came to be added a whole new dimension to the magnificent intellectual forum. It was the participation of numerous other thinkers, brought to the venue by the modern technology of telecommunications. The main speakers were six, but the electronic participants were a score of scientists who were teleconferencing in from remote areas of space-time.

Other Conference Participants

By using modern teleconferencing screens that could be activated by the three "A-Team" philosophers (Aristotle, Augustine and Aquinas), there were panelists available on every conceivable topic.

On physics, there was Newton and there was Einstein, plus a third, round-faced, jolly-looking Belgian priest (Georges Lemaitre) whom many might not recognize, but who they understood was the theoretical discoverer of not only the Big Bang, but the "inflationary" moment after it. The philosophical implications of those two events, which we will explain later, were clear and convincing to the A-Team.

On anthropology, there was some consternation, when Darwin's name was suggested. The three A's discussed briefly the problem of dealing with a theory which had convincing elements as to common ancestry and chronological development, but seemed to violate the Second Law of Thermodynamics.

A call was put in to Leonardo DaVinci, and he confessed to be equally flustered. This surprised them, as they assumed he was the most modern polymath, and should know how to reconcile the principles of physics with the arguments of evolutionary biology. (The Second Law of

Thermodynamics, which dictates that all things tend naturally towards disorder – thus seeming to contradict Darwinian natural selection – was agreed by all to be the most important law in all of science, without which even time cannot be understood.)

How can the Second Law be reconciled with Darwinian evolution? - they asked. After talking to Leonardo, calls were made to Benjamin Franklin and even John Forbes Nash (he of the "Beautiful Mind") in the hope of finding someone who might provide a contemporary understanding of consensus anthropology (what journalists nowadays call "settled science"). Both of them demurred, but Nash suggested another, more contemporary name: Buckminster Fuller.

"Bucky" immediately suggested a solid panel on modern, post-Darwinian biology. "Use a trio of eminent scientists who modified Darwinism by the more advanced concept of 'Punctuated Equilibrium,' which admits that there are big jumps or 'saltations' in the historical, fossil record, instead of a smooth ramp or slope towards more organized species." And, continued Buckminster, "the trio of Stephan Jay Gould, Lyn Margulis and Niles Eldredge have more or less convinced everyone that the climb up the evolutionary ladder was more like a staircase. Go with those three, who have sterling credentials."

And he added, as an afterthought: "Be sure to have handy the 28 volumes of the *Traite de Zoologie* published by the French Academy of Sciences under the leadership of its chair, the eminent Dr. Pierre-Paul Grasse. (Dr. Grasse had forcefully argued that organic life does not obey the Second Law, because it seems to have an internal motor that produces greater order over time.)

Whatever the merits of the argument by Grasse, it was clear that genetic roots were found deep into the history of animals – arguably going back to the very first amoebas (single-celled combinations of molecules that began to appear and to replicate themselves about four billion years ago). And in that replication, which was initially asexual and later gloriously sexual, all traits seemed to be inherited, by wonderful and yet somewhat counterintuitive rules discovered by an Augustinian monk (Gregor Mendel).

And so they had asked Mendel, as the precursor of that science, to be on the line from his monastery, where he had continued experimenting with peas and other organisms for a couple of hundred years now.

They also had great mathematicians and astrophysicists on standby, including Euclid, Archimedes, Pythagoras, Euler, Copernicus, Leybniz, Descartes, Kepler, Galileo, Stephen Hawking and Brian Greene.

On economics, they decided on Adam Smith, of course, and added Keynes and Hayek to reflect the tension between laissez-faire and government control of investments.

Just as they were ready to get started, Bucky called back with another suggestion. "Listen, guys," he said, "if you get stuck on any complicated matters of modern metaphysics, there is someone that should be on the line, listening and ready to carry the argument. His name is Ludwig Wittgenstein, as in *Wittgenstein's Poker.*" Hearing no response, he reassured them: "The reference to 'poker' is not to the game, but to a famous incident in which he and his Austrian friends almost went at it with a hot poker...."

The Rules on the Consideration of "Sacred" Texts

On this issue, there was a lot more consensus than one might have imagined; and here again the referee was Augustine, who wanted to assure Aristotle that he would not be overwhelmed by the other two members of the A-team, who were strong believers.

"In case you never saw my pronouncement, which Aquinas obviously accepts, the general principle *when there is apparent contradiction between science and scripture is that we must be misinterpreting scripture.* Reason and faith go hand in hand. One enlightens the other; neither detracts from the other."

But Aquinas wanted to give Aristotle greater reassurance. In this forum, which is scientific, we only consider two ideas that flow from the bible. One is the idea that things come in three dimensions; for example, he said, "the human mind is made up of three elements: memory, understanding and will."

"Well said," replied Aristotle, "that mirrors my own definition that there are three substances or primal elements in nature: the physical, the mental and the spiritual. It would be foolish to leave out any insights that come from texts the people consider 'sacred,' at least to the extent that they have led to more civilized behavior."

Aquinas piped in, once again, observing that Jacques Maritain had argued, on that very point, that the natural law seemed to flow from divine law, and that there was an "ontological" (inherent) as well as a "gnoseological" (learned) component to that ascertainment.

At this point, Augustine interrupted: "Gentlemen, there is a limit to the ability of common folks like the rest of us to reach your level of analysis. Let's just say that we don't reject whatever insight, in supposedly 'revealed' texts, does not contradict reason."

There followed a brief discussion about the matter of free will, which they sensed had become a bit controversial in recent times – with the advent of biological determinism (also known as "Behaviorism"). But both Aquinas and Aristotle quickly agreed that our species is distinguished from the animals by an evident capacity to choose between what moderns call "social" and "antisocial" behavior and what ancients referred to as virtuous, versus un-virtuous conduct.

These things were self-evident to them. But there was another important principle to discuss, before delving into the full-fledged analysis.

The other important principle was left to be a conclusion, rather than a premise, though to them it was self-evident. It is the realization that humanity has been unable, over time, to manage its own instincts without at least one, additional intervention by the infinite (or near-infinite) being that propelled the Big Bang and organized the universe during the inflationary moment so precisely that it could still function well after **13.7 billion years**.

Old and New Testament believers referred to the matter as "salvation" or "redemption," meaning a particular point of time in which the Big Banger seemed to have intruded into human history.

But this was not an *a priori* assumption. This was to be argued and proven, to the best of their abilities. But first let's clarify what we mean by "proving" something.

Methodology for "Proving"

For reasons that will become clearer a little later in this book, it is not easy to say, even using basic, empirical tools, such as measuring devices, that something is absolutely, irrefutably, unquestionably true. Or, as previously mentioned, what is nowadays referred to as "settled science."

I don't want to get too deeply into the philosophy of science. In other books I have written, I have delved into what constitutes scientific "certainty" or a "fact" in the sense that there is really no rational doubt about it.

The most commonly accepted definition of a scientific proof is something that can be unverified ("falsified") by testing. Dropping an apple repeatedly to show that it falls due to gravity is the kind of test that can verify, or falsify, a theory. If done in a vacuum (such that there is no air resistance), the exact value of "g" (the gravitational constant) can be disproven (falsified) with a high degree of precision. (Karl Popper is credited with this definition of science; a book to read is *The Logic of Scientific Discovery*.)

Perhaps the most celebrated instance of a very complicated theory that could be unverified with fairly simple testing was Einstein's theory of relativity. It says that light bends around any object, due to the same gravitational force that attracts the apple when it falls. To un-verify it, it is necessary to find a ray of light that passes very close to a very big object (like the sun). However, in order to be able to see the light coming from the distant object (such as a star) that sends it, we need to have a dark sun, otherwise we can't see the light coming at us from the distant star.

As it turns out, that can only be done when there is a total solar eclipse, which doesn't happen all that often and, when it happens, is not easy to see from any old place on the face of the earth. The best place and time to see it was in 1922, from Australia; and there went no less than ten teams of scientists from a handful of countries. Only a couple were able to take the pictures at just the right time.

The pictures they took proved Einstein's equations of relativity; and, since that time, the equations have been used for all kinds of practical things, such as to predict the flight of rockets and to adjust the exact path and speed of electromagnetic transmissions, such as those which allow me to send messages on this computer to the other side of the world.

There are other categories of "proofs." The so-called "bell curve" is a statistical compilation of things like human intelligence, height and maximum age. It is not a strict cause-and-effect proof, as gravity or relativity, but it is pretty darn predictable.

DNA tests in criminal trials are used to prove the likelihood that someone's identity is matched with blood samples found on the victim. (In the famous O.J. Simpson trial, the bloody glove found at the scene contained blood whose DNA was matched to Simpson's with a probability of 150 million-to-one.)

I guess that's pretty certain, although it was not enough to convince the jury – perhaps because they thought someone may have planted the bloody glove.

To a great extent, statistics is the methodology of the social sciences, among which are psychology, sociology, criminology, anthropology and economics. Economic theories are particularly difficult to prove, even with computers that can crunch enormous amounts of statistics. One reason is very similar to the uncertainty which affects the science of meteorology (weather forecasting). There are just too many factors that determine the future temperature, wind velocity and moisture in any particular spot on the earth.

In the case of sciences that deal with human beings, there is a second, important reason for the uncertainty – though it is not one accepted by certain modern scientists. It is the fact that human beings are inherently unpredictable.

I will discuss the arguments pro and con the existence of free will later on. Ultimately, the reader will have to come to his/her own conclusion on that issue. On that, as well as other very basic concepts, we must occasionally have to just assume what our intuition tells us.

Here's why.

If Nothing Is Assumed, Nothing Can Be Proven

As C.S. Lewis said, "if nothing is assumed, nothing can be proven." Some things just have to be assumed. If you don't believe that there is a difference between what is true and what is false, then it is difficult to continue any rational discussion. Or if you are not sure that there is a reality which often contradicts what we understand to be "illusions" that intrude occasionally into stable personalities and constantly into pathological ones, then you can stop reading right now.

Other things are provable, but not with the kind of certainty that we extract from experiments that yield measurements and can be repeated (the methodology of the "empirical sciences"). For example, the social sciences often have to rely on statistical correlations that indicate a high degree of scientific probability to explain some phenomenon. (In criminal law, we refer to those proofs as being "beyond a *reasonable* doubt.")

A third category is things that are more probable than not. They are not "settled science" in the sense of being accepted as "fact" by the great majority of savants. In this category are things like the desirability of monogamy, the undesirability of incest, and similar prescriptions of what philosophers call the "natural law."

The natural law is a fairly established set of norms that guide societies in the modern era. They include the notion of equality and the general idea that "might does not make right." In other words, that the law of the jungle does not apply to human beings. (More on the natural law in later chapters.)

The last category of provability is one that I will call "consistent with reality." This category includes many insights that come from allegedly revealed texts (e.g., the bible) and that are not provable or testable but are what we lawyers call "consistent with the facts." For example, the idea that humans are more than just matter and intelligence – that humans seem to have an internal, moral barometer (conscience) is ultimately going to be accepted in this book as more than just a possibility or hypothesis.

In the chapters that follow, I will try to prove that the existence of a conscience or inner-voice in humans is not only scientifically probable, but consistent with a belief system (the Judeo-Christian one) that has changed the course of history.

But that is for my reader to decide. And it is, in the main, what this book is about.

And now it is time to start with what science suggests is the beginning of time itself.

Chapter I

IN THE BEGINNING: ENTROPY

Just over a billion years ago, many millions of galaxies from here, a pair of black holes collided. They had been circling each other for aeons in a sort of mating dance, gathering pace with each orbit, hurtling closer and closer. By the time they were a few hundred miles apart, they were whipping around at nearly the speed of light, releasing great shudders of gravitational energy. Nicola Twilley, *New Yorker Magazine,* February 11, 2016.

The scientific discoveries of the last half of the twentieth century make it hard for anyone to be a non-believer. The prior century had shown us science as its most precise, predictive self. In order to find out how things came to be, all we had to do was dig a little bit, break up what we found, lay the pieces on the laboratory table, and re-assemble them with a smattering of intellectual glue.

By the second half of the twentieth century, a famous writer was proclaiming "the end of science." Other scientists predicted that humans would soon produce super-computers that could do our thinking for us, our breeding for us, even the raising of our kids.

Everything we needed to know would soon be found in properly calibrated equations. Everything we needed to survive would soon be produced by computer-designed machines and robots. Physics and math (and the other empirical sciences) would soon kick psychology, philosophy and theology out of the campus.

No more humanities! Computers and those who design them are now in charge of the universe.

Then came the startling scientific discoveries of the second half of the twentieth century, including the discovery that one moment there was total darkness and timelessness and the next moment (in one tiny fraction of a second) astounding mass, energy and organization of the universe, for which not only was there no known cause, but for which there could never be, empirically speaking, a precise explanation.

Just when empirical science was about to evict the arts and the humanities from the campus, the whole idea that everything would ultimately be a matter of experiment and measurement (the whole edifice of scientific certainty) came tumbling down.

But that came a little later. In the first couple of decades of the twentieth century enlightened minds concluded that theology was something obscure and needed only as a stopgap measure, while science caught up with a clear and concrete explanation. Intelligent, well-read people concluded that theology was a way to explain small gaps in our scientific understanding. (As we shall see, that notion came to be known as "God of the Gaps.")

Three phenomena brought down the edifice of scientific certainty: they were the Big Bang, the Inflationary Moment, and Dark Energy.

Because I don't want to scare away my reader, I will only elaborate here on the Big Bang and only delve lightly into the Inflationary Moment. (Dark Energy is really just a cryptic name given to a rather immense phenomenon, constituting about 70% of all the mass and energy in the universes. For a homespun discussion of what is an extraordinarily complex question, the reader is referred to my book: *On the Likely Origin of Species.*)

As for what happened at the beginning of the universe and immediately afterwards, it is important to our analysis for the obvious

Science and the Theory of God

reason: To know where we are going as a species we need to know where we came from and how we were made.

Not to mention who made us.

In the second half of the twentieth century, those questions were answered in a way that made it difficult to be a non-believer. The last half of the twentieth century, as we shall see, was a bad time for atheists.

Let me explain.

The Day Without Yesterday

For about half a century now, we have all taken for granted that some sort of explosive event lit a fuse in the heavens and propelled clusters of stars into orbits, organizing them into what we call galaxies. But those are all relatively recent discoveries. *Throughout most of the modern era, scientists and other savants have mostly assumed that the universe was infinite – that it had no beginning.*

Those who believed that the physical universe had a beginning assumed that there had come a time when the infinite being (the biblical creator and the philosopher's "First Cause") had simply decided to create material beings. But it was mere speculation, fueled to a great extent by allegedly revealed texts in something called the "scriptures," which was a term applied to both the Hebrew bible and, more recently, the Christian one.

All that speculation (and all that reliance on biblical accounts) ended about three quarters of a century ago. It was then that the best and brightest of our species confirmed, using the scientific method, that the universe did, indeed, start with a Big Bang. Matter, time, space and order appeared almost simultaneously in a spectacular, creative burst. (The book to read is by John Farrell and its title is the one I gave this section: *The Day Without Yesterday*.)

Not that such discoveries solve the issue of "why." For centuries prior to being able to pinpoint the beginning, as well as the startling idea that time did not exist before the Big Bang, the classic discussion among savants entailed the logical quandary: If there is a god, and if he/she is infinite in time, what prompted him/her to create the universe? And the related question: Why did the universe begin at the particular moment that we now estimate happened 13.7 billion years ago?

Leibniz stated it thus: "Why, if He has endured through an infinite time prior to creation, did not God create the world sooner?"

Why, indeed?

Immanuel Kant thought the entire question absurd. He argued that "the world has no beginning" but "is infinite as regards…time…" Watch how he reasoned:

> *Since the beginning is an existence which is preceded by a time in which the thing is not, there must have been a preceding time in which the world was not, i.e. an empty time. Now no coming to be of a thing is possible in an empty time, because no part of such a time possesses, as compared to any other, a distinguishing condition of existence rather than of nonexistence; and this applies whether the thing is supposed to arise of itself or through some other cause. In the world many series of things can, indeed, begin; but the world itself cannot have a beginning, and is therefore infinite in respect of past time.*

I have never been a fan of Immanuel Kant; after reading that quote, my reader might understand why. On this issue, I prefer to quote Isaac Newton, who was more conventional in his view that the universe was created by God *ex nihilo* (out of nothing). He also believed that prior to the beginning of the universe there was an infinite duration devoid of all physical events.

Here's how William Lane Craig describes the rest of Newton's vision of what was happening before the commencement of all things, which he describes as a "beginningless" time, in which –

> *at some point a finite time ago the universe came into being. For Newton our familiar clock time is but a 'sensible measure' of this absolute time, which, he says, 'of itself, and*

from its own nature, flows equably without relation to anything external, and by another name is called duration.'

Newton's vision (based almost totally on deductive reasoning and biblical beliefs) turned out to be pretty accurate. It took two discoveries to get us to what we now call the Big Bang, which is fairly settled science. What might surprise my readers is that the Big Bang was really two events.

The Beginning of Time: Two Big Bangs

As we shall see, the Big Bang was really two bangs. The first one is the one most mentioned and in many ways most startling. The story begins with a Belgian monk, by the name of Georges Lemaitre.

Lemaitre turns out to have been the chief proponent of both the first Big Bang and the "second bang"– to which we will refer soon. As to the first bang, the one that simply proposed that the entire universe began at one, single, infinitesimally tiny spot or "singularity," Lemaitre had a hard time convincing elder statesmen like Albert Einstein and Sir Fred Hoyle.

Like most scientists, Einstein and Hoyle were advocates of a model that argued for a "steady-state" universe. This model resembles a beating heart: a pulsating, cyclical, shrinking-and-expanding mass of particles that did not have a clear beginning or a predictable end.

Lemaitre argued with equations; others later argued with empirical proofs. One proof was accidental; it happened when technicians picked up very, very small radio static of the kind that bothers us when we are trying to listen to music. This noise, which is something that we cannot eliminate, is the result of a very distant explosion whose magnitude, when divided by the volume of the allegedly expanding cosmos, is the equivalent of a couple of degrees of heat.

A couple of American radio technicians (Arno Penzias and Robert Wilson) managed to confirm that a temperature slightly more than two degrees above absolute zero (which is very, very cold) is found everywhere our instruments are able to measure temperature in the night sky.

This discovery constituted what turned out to be the second and confirming proof that there was a Big Bang and that it took place at a single point in time and space. Another proof came from a less accidental discovery, which came to be called the "red shift."

The Red Shift

It sounds complicated but it really is quite simple and intuitive. Light waves, which travel in a vacuum, are much like sound waves, which travel through the atmosphere. They both travel as waves of a certain size. When light or sound travel towards you, the waves become compressed by the speed of their approach; when they travel away from you, the waves stretch or expand in size.

We all know the phenomenon of sound waves from an approaching automobile or motorcycle, which sound quite different from the same vehicle when it passes us and rapidly moves away. The reason is that the sound from shorter wavelengths is different from the sound made by larger wavelengths.

Light waves do exactly the same thing, except the difference is not picked up by our ears, but is picked up by a spectrometer – a prism-like machine that breaks up the light so that its color is seen from its wavelength. Stars like the sun are luminous because they have burning gases; and the color they emit is a bright red.

As soon as scientists began to analyze the tone of red from various stars, they realized that it varied according to the distance of the star: the farther the star was, the more the red color "shifted" in the spectrograph machine, indicating a higher speed.

Thus it happened that Lemaitre's theory now had two proofs; and many more kept coming.

There are other proofs for the expanding universe; scientists in various fields have even confirmed, to a fairly precise degree, the age of the universe, from that first big explosion. They are pretty sure it all began 13.7 billion years ago.

That's all well and good. But how exactly did an explosion that scattered particles all over the place, at enormous speeds and with an electric charge (negative in the case of electrons, positive in the

Science and the Theory of God

case of protons) end up coalescing into nice galaxies, full of suns and planets?

The explanation is so startling and so unexpected that it makes an engineer like me, or anyone with any kind of mechanical curiosity, marvel at the Big Banger. Scientists call it the "inflationary moment."

(Inflation in this case is not a reference to the value of money, but to a remarkably fast and powerful expansion of the singularity from a single point in space and time to a little, organized, mini-universe.)

It is the most underrated, most surprising, least understood and least-likely-to-ever-be-understood mystery of science. But it has very well calculated dimensions. It happened in a small space which you can call "nature's little workshop" or, for believers, "God's little workshop."

The Inflationary Moment: Nature's Little Workshop

It is not easy to understand or explain what happened in the first moment after the Big Bang. I have to thank a Princeton-based theoretical physicist named Brian Greene for what little I am able to grasp of this short period in the history of our universe. The good professor has written various books which are readable by amateur physicists like yours truly. (One that is particularly readable is *The Fabric of the Universe*.)

The gist of his explanation goes back to the Bill Paley argument for the existence of God based on the design of a watch that was accidentally dropped and is found in the bushes next to a well-worn path. The watch is a small, intricate machine that clearly reflects the workings of a smart craftsman.

Let's listen as Paley explains his memorable metaphor in support of some sort of creation:

> *In crossing a heath, suppose I pitched my foot against a stone, and were asked how the stone came to be there; I might possibly answer, that, for anything I knew to the contrary, it had lain there forever: nor would it perhaps be very easy to show the absurdity of this answer. But suppose I had found a watch upon the ground, and it should be*

> *inquired how the watch happened to be in that place; I should hardly think of the answer I had before given, that for anything I knew, the watch might have always been there....There must have existed, at some time, and at some place or other, an artificer or artificers, who formed [the watch] for the purpose which we find it actually to answer; who comprehended its construction, and designed its use.*

Borrowing the Paley analogy, Brian Greene describes the first mini-universe as a little clock that is first assembled and then wound up to start ticking for about fourteen billion years. (Due to increasing entropy, which means disorder, the universe and all its matter is disintegrating slowly. We will explain that shortly.)

The speed and power with which the little clock was assembled was calculated by a couple of scientists named Alan Guth and Andre Linde (the former from M.I.T. and the latter at Stanford). The numbers are scintillating, extreme, as close to infinite as any phenomenon known to scientists.

You want to know how long it took for the little assembly to come together? Try a fraction of a second with the number one at the top and ten with 36 zeroes underneath. You want to know how big was what Greene calls the inorganic "stitch" in the "fabric" of the universe? Try a fraction of one meter with one at the top and ten with 42 zeroes at the bottom. In other words, extremely small in space and extremely fast in time.

But not everything was small. In fact, the mass that appeared and the energy that was applied to create space for that mass to expand were both enormous. Under the law of conservation of energy, no new mass or energy came into existence after that fraction of a second. So everything we see in the universe now was contained there, in a tiny package that was as complex as it was small.

It was also *enormous* in terms of its mass (which we commonly refer to as its weight); *now that was extraordinarily large.* How it fit into that tiny, little space is beyond me. I try not to think about it too much, though perhaps it is necessary if one wants to try to understand what person or entity made it happen.

Let's accept the notion that our minds must be stretched, much like the stretching that took place at the very beginning, when not only did we see mass and energy take shape, but a complete "witch's brew" of elements, heat, radiation, and complex forces pulling and pushing in every direction.

That is remarkable enough: One moment there was nothing; the next there was a mini-universe of components, interacting like an ultra-fast pinball machine. But what is more remarkable is that the mini-universe was completed to such a degree of perfection that, so far as we can tell, not one new component was added for the next fourteen billion years, and yet the original formula somehow produced all the galaxies in the night sky and at least one planet with glorious, biological organisms possessed of names like Madonna and LeBron James.

In other words, the mini-universe that came together at the beginning of time was so well planned that its components could deteriorate and disintegrate from that point forward and still come together in a magnificently well-oiled biosphere we call the planet Earth. This phenomenon has to be further analyzed, for its philosophical implications.

Like it or not, we must discuss entropy.

The Unbearable Strangeness of Entropy

I promise my reader that this is the last discussion of an esoteric concept derived from thermodynamics or physics or similar empirical science.

The simple reason we must deal with entropy is that it happens to be easily the most philosophically relevant physical law of all. Technically, entropy is a measure of the randomness of things — as opposed to the organized tidiness that is exemplified by well-crafted machines, such as Paley's watch or Greene's clock. Entropy dictates that eggs which break do not usually come together on their own; it means that old cars rust if not cleaned and oiled; and that the gas inside a bottle of soda that is left open leaves the container and never returns.

Under something called the "Second Law" (of thermodynamics), entropy works to make things less organized; it is relentless in that quest. Absent an outside force that works on any closed system, it will ultimately disintegrate into the most elementary particles.

So that begs the question: If matter was at its highest level of organization at the end of the inflationary moment, and if it has been breaking up, rusting, decomposing for almost fourteen billion years since then, how can things have come together so nicely in the end?

I spent a few years studying that question – reading what theoretical physicists and astronomers and geologists said about the issue. In the end, I concluded (in a book I called "On the Likely Origin of Species") that the sheer amount of order (reverse entropy) was so high at the beginning of time that it could slowly wear away and still pack enough wallop in one corner of the universe to form a planet as wonderfully interacting and well designed as ours.

All of that could have happened spontaneously. What is harder to accept is the notion that organic life on earth, and in particular human life, also happened spontaneously.

The question of whether biogenesis (the beginning of life on earth) happened spontaneously is not one that I can tackle here. Suffice it to say that scientists have been struggling to replicate the creation of life from inorganic chemicals for three quarters of a century, without much success. Not only that, but some eminent scientists are convinced that organic life does not follow the Second Law – that it beats to the rhythm of a different drummer.

I won't even try to settle that argument here. Organic life may or may not have happened spontaneously. It may or may not have evolved spontaneously from amoebas, four billion years ago, to our species. That is a topic for another book – and is a key component of my previously mentioned effort: *On the Likely Origin of Species.*

But here we must delve into what science says about the beginning of **human life**.

But before we go there, let's summarize what happened from the beginning to approximately fifty thousand years ago.

Science and the Theory of God

From the Big Bang to the First of Our Species

Popular science writers (and journalists and even some academicians) love to give us their most imaginative metrics that describe how long it has taken to go from nothingness to the first human couple, whom they love to call "Adam" and "Eve." (The book to read is: *In the Footsteps of Eve*, by Lee Berger).

One thing that scientists all agree on is that fully intelligent beings like us only walked the planet for the last 50,000 years. If you compare that to the age of the universe, at almost fourteen billion years, and use the age of humans as a factor of the universe's age, the arrival of *homo sapiens* is the equivalent of just a few minutes.

Scientists, using those metrics, are mesmerized by the relative youth of our species. It's almost as if they are trying to convince themselves (and us lay people) that because it took so long to get from elementary particles to mammals, the appearance of humans in recent times is somehow more *predictable and more devoid of mystery.*

The truth is exactly to the contrary. The appearance of humans on this planet is such a scientific break from what came before that it begs for explanation in terms of what philosophers and theologians call a "special creation."

A special creation is no more and no less than another "Big Bang" – another moment in which there is no plausible explanation for what appeared in nature, based on what preceded that phenomenon.

Think about it. There is now scientific consensus that humans acquired cognitive faculties perhaps around fifty thousand years ago. But scientists admit that they have no idea of how it happened. A typical description of the chronology is found in a recent *New Yorker Magazine* article; it casually explained that: "About fifty thousand years ago... homo sapiens were beginning to replace our Neanderthal cousins as the planet's dominant species of ape."

So the chronology is agreed upon. It happened about 50,000 years ago. What is not clear is how it happened or where it happened, or which particular pre-human somehow acquired the power to think in abstract terms or to paint nice pictures in caves.

(The notion that it all happened 50,000 years ago might not even be a reliable chronology of the advent of our species. I will leave that discussion for another day. For now, let's just assume that chronology, since it seems to reflect the consensus of anthropologists.)

In the old days, we would say that there was a "missing link" between our cousins, the big apes, and us. Nowadays both academia and the popular media skip that whole issue and pronounce us as first cousins of the Neanderthals, as well as second cousins of the African "Australopithecus," better known as "Lucy."

You cannot blame the popular press for being less than precise: The reason that the popular literature does not explain the arrival of humans on the planet is that there is not much hard science on the subject.

Here is a typical, puzzling passage from an otherwise solid scholar, Yuval Noah Harari, an Oxford Ph.D. who currently teaches at the Hebrew University at Jerusalem; it comes from his recently published (2011) and much-celebrated book entitled, *Sapiens, A Brief History of Humankind*:

> *On a hike in East Africa 2 million years ago, you might well have encountered a familiar cast of human characters: anxious mothers cuddling their babies and clutches of carefree children playing in the mud; temperamental youths chafing against the dictates of society and weary elders who just wanted to be left in peace; chest-thumping machos trying to impress the local beauty and wise old matriarchs who had already seen it all. These archaic humans loved, played, formed close friendships and competed for status and power – but so did chimpanzees, baboons and elephants.*

It cannot be good science to argue that big apes, living two million years ago in East Africa and looking like us, in the sense of walking upright, were humans just because they had alpha males and anxious mothers "cuddling their babies…"

But it is refreshing to see Harari admit that their habits and traits were no different from those of modern chimpanzees, baboons and elephants.

From Apes to Men in Four "Saltations"

Honest scholars, who have thoroughly studied the matter, suggest no less than four major (and mostly unexplained) evolutionary jumps from the big apes who roamed East Africa two million years ago to modern humans. The last jump (called "saltation" by scientists, from the Latin word meaning "jump") is the one that they estimate took place in the last 50,000 years.

That saltation is described by Stanford anthropologist Richard Klein as consisting of a "genetic mutation that promoted the fully modern brain." Interestingly, he admits that this neurological metamorphosis "cannot be tested...by experiments or by analysis of relevant human fossils."

Regardless of how this last quantum-leap happened, Klein and the rest of the evolutionary biology establishment are happy to conclude that humans came to possess "the full blown capacity for culture, based on an almost infinite ability to innovate." The newly acquired trait, unique in the animal kingdom, allows humans to "conceive and model complex natural and social circumstances entirely within our minds."

Following Klein's dictum, which allots to humans an "almost infinite" ability to innovate and assuming that there is a creator, who has, by definition, an "infinite" ability to innovate, we see the logical likelihood that humans are the image and likeness of their creator.

But we're getting ahead of ourselves. All we know up to this point is that a Big Banger made a mini-universe out of nothing; that the mini-universe had such fine-tuned components that it expanded for 14 billion years into marvelously interacting galaxies, at least one of which has a marvelous biosphere we call earth; and that on that earth, perhaps 50,000 years ago, an organic being evolved to the point that she and her descendants possess an "almost infinite" ability to innovate.

So let's see what we can conclude from the above facts, and from the rest of what Harari calls "A Brief History of Humankind," Stephen Hawking calls "A Brief History of Time" and Bill Bryson calls "A Short History of Nearly Everything."

In the next section I will tell you what I conclude, using reason and scientific methodology only (no biblical texts).

From Mini-Universe to Humankind

The analysis begins, necessarily, after the Big Bang. We can only speculate as to what happened before it; from what we do know, there was nothing before: no matter, no energy, no order – not even time.

Someone or something caused the mini-universe to be formed. Was it an infinite being? Was it an all-loving being?

Let's take those questions one at a time. Whatever or whoever made the mini-universe come to life in a fraction of a second had at least two important features: (1) he/she/it had at least near-infinite intelligence and (2) he/she/it was "timeless" in the sense of existing before time was invented.

A timeless being, with intelligence so great that he/she/it could create a mini-universe out of nothing, propelling it forward for 14 billion years in such well-planned motion that Albert Einstein (with an "almost infinite" ability to innovate) could evolve from the enormous, and enormously complicated mixture of elements produced at the Big Bang sounds very much like an infinite being.

Our instinct quickly makes what I call a "leap of logic," and concludes that it was, indeed, an infinite being. It is not a "leap of faith" based on some biblical text. It is more basic than that.

And there is some interesting math to support that leap of logic. Let me take the simplest of all, rather than delving into Einstein's equations of relativity, which, parenthetically, collapse into infinite sums as we extrapolate back in time to the Big Bang.

At the Big Bang, and for an exquisite moment in time, there was an enormous amount of mass occupying a single point in space. Mathematically, *that means infinite density*.

Period. No "ands, ifs or buts" about it. Infinite density, following timelessness and emptiness. Could any force or intelligence, other than an infinite one, have created it? Let me restate it: One moment there was no time and no mass; the next there was time and infinitely dense mass.

Too bad Thomas Aquinas is not alive or he would frame yet another proof of the existence of God – this on strictly mathematical grounds.

Now let's talk about love.

Does God Have to Be All-loving?

There is no scientific proof that God is all-loving. It's just common sense.

Let's say you are an infinite, timeless, all-knowing being, sitting around with nothing to do other than exist. (For the moment, we assume that you, the infinite being, are one entity and not a trio of entities; later we will delve into the notion that God is a community, rather than a lonely individual.)

If you are infinite, timeless and all-knowing, you are probably pretty satisfied with yourself. Who wouldn't be? It is fair to assume that you have no worries. You don't need to struggle for survival. *You sure don't have any problems of self-esteem.* You don't need anybody, other than yourself.

So what do you do for recreation? Well, here we have a quandary. Scientists struggle with the question of whether humans need recreation, like music and dancing and theater and the arts generally. Some rather dour scientists, who are convinced that humans have but one overriding instinct (to survive long enough to procreate) struggle with the whimsical, and the purely artistic traits in our species.

We are smarter than that. We know humans need to hear the waves of the ocean, the sounds of a harmonious tune, the voice of a child who says "daddy, mommy" or the whisper of a lover who says; "come closer and touch me and kiss me."

But does God?

I really have no answer to that. I tend to think God is infinitely self-sufficient, that he/she needs nothing, and could have spent all of eternity without creating any material beings. But it really doesn't matter. The point is that he/she did.

And in all likelihood, with a fair degree of scientific certainty, the act of creation was motivated by nothing but unbounded love. Why do

I say that? Because it's the only logical explanation for an infinite being to create a finite being.

Many humans claim that they feel the love of God; many have given their lives to God in what they consider to be a reciprocal act of love. Social scientists – in particular those dealing with the treatment of addiction to alcohol and other substances – assure us that the addict needs to sense God's love in order to take the final step to sobriety. Those, I suppose, are all solid evidence that God is a loving entity.

For me, it is pure and simple logic. An infinite being is almost certainly emotionally self-sufficient. And the more emotionally self-sufficient one is, the more one is able to love. The two things seem to go together.

The reader might conjecture that I am letting an emotional-religious bias take over – in particular the biblical images of God as a benevolent parent that watches over us and wants us to be united with him/her at the end of our earthly lives. Anyone who was lacking in parental love might well turn to an ethereal being, powerful and kind and parental, who might supply the love missing in one's own upbringing.

It doesn't work in my case. I was loved dearly and tenderly by both of my parents. I have never lacked for love from elders and peers alike, including and most notably my wife of forty years. I don't emotionally need the love of a creator to feel whole.

For me, the benevolence of God becomes evident when I observe the beauty of this earth, of creatures big and small, of babies and beautiful actors and actresses, of singers like Toni Braxton and Frank Sinatra, of playwrights like Shakespeare and Cervantes, of poets like Robert Frost and Gustavo Adolfo Becquer, of writers with flowing prose like Churchill and Manchester and Jose Marti. When I experience the beauty of life, the joy of family and friends, I am compelled to conclude that God is all-goodness and all-loving.

And that begs the question of why things went awry. Why does this particular species – our species – need redemption? And how did God, in his/her infinite wisdom, let that happen?

Chapter II

OF ANGELS, ANIMALS AND MEN

God made the angels to show Him splendor, as He made animals for innocence and plants for their simplicity. But Man He made to serve Him wittily, in the tangle of his mind. Sir Thomas More, in *A Man for All Seasons,* by Robert Bolt.

I was favorably impressed (though not necessarily convinced) when I first read the above line from the famous play, *A Man for All Seasons,* which later became an Academy Award winning movie. Robert Bolt, in one short phrase, captures what might well be the motivation behind the entire hierarchy of creation, from angels to animals and plants, to man.

We don't really have any scientific proof of the existence of angels, but if they did exist, I suppose they would be purveyors of splendor that would please the creator.

As to animals, plants and humans, we have absolute proof of their existence. Suggesting that God made animals "for innocence" and plants "for their simplicity" is a nice, plausible proposition. The phrase gains in persuasive power when completed with the more profound insight: that man was created to "serve [God] wittily, in the tangle of his mind."

Humans have certainly had a tangled relation with their perceived creator. From the earliest indications of recorded history, men and women have used their wits to try to communicate with a more powerful and primordial presence out there, as reflected in the oral and written histories of most civilizations for which we have records.

Bolt's description of man's attempt to relate to God "in the tangle of his mind" is historically accurate: Religious beliefs and practices, as can be traced back to the earliest civilizations, reflect a tangle of ideas, a confusing and often contradictory attempt to understand a transcendent reality. By my modest reckoning (and I am not even remotely a religious historian), the tangle of ideas culminated, with far greatest impact on humankind, in the so-called "scriptures," i.e., the Torah (Hebrew Bible) and the Christian Gospels and related writings that claim to have fulfilled it.

The Book of Genesis describes, in marvelous detail, the fall of humankind from a paradise-like lifestyle, to blood, sweat and tears.

An Alternative Theory: Scientific Materialism or "Reductionism"

Before we consider a theological model of fallen man, let's review what modern "reductionist" scientists believe. I use the term "reductionist" (whose opposite is "holistic" or "integral") to describe a modern school of scientists who espouse the theory that humans are really slightly smarter apes. Those scientists deny that humans fell out of step with the creator; what's more, they hold to the view that there was no special creation at all – no need for any intervention from outside the material universe needed to transform an animal into a human.

Because they reduce everything to mechanistic notions of evolution, the term "reductionist" applies. They are also referred to as scientific "materialists" because they are convinced that all beings, including our species, are composed of one substance: matter.

Scientific materialists do not believe in a "mind-psyche" that is separate from the brain. They do not believe in a conscience, soul, or a free will separate from the appetites that drive humans to act in a particular way. They deny the existence of any transcendent being.

I like to say that for them "only matter matters." (M. Scott Peck refers to it by the modern saying: "What you see is what you get.")

Scientific reductionists or materialists are convinced that religion is a fabrication – an artifice, if you will, by humans to explain holes in the scientific puzzle. Because such fabrications are not real explanations, they argue, religious explanations are like dental fillings used temporarily but quickly discarded when the permanent, solid, enamel crown replaces it.

Scientific materialists have coined a term for their all-embracing, mechanical reductionism, which sees the "God Theory" as an interim, irrational solution that will ultimately be discarded entirely, as empirical science solves all the riddles in humanity's search for truth.

They refer to the God Theory as "God-of-the-gaps."

"God of the Gaps"

The idea that mysterious phenomena, for which no scientific explanation could be found, must be consigned to the realm of religion (i.e. to the realm of the "occult") has been referred to as "God-of-the-gaps." It's a fairly descriptive term, since it fills in gaps in our understanding of reality with the notion of an all-powerful "God," who simply dictates some reality that cannot otherwise be explained.

Throughout recorded history all kinds of tragedies – as well as felicitous happenings – have been ascribed to divine intervention. If they were good things, such as manna (bread) falling from heaven, they were called miracles. If they were bad things, such as epidemics and unprecedented bad weather (e.g., a flood), they were thought to be curses from above.

Much has been accomplished during the last half millennium to explain away such mysteries, whether good or bad. Perhaps we should have expected that science would become a bit presumptuous.

Arguably, the high point of scientific over-reach was the midpoint of the twentieth century, when scientists found a way to peer into the *microscopic world* of the atom and into the *macroscopic world* of distant galaxies. For a while there, it seemed as if nuclear physics would merge into organic chemistry, and organic chemistry would explain biology.

Science has, indeed, closed a lot of gaps on the way from the Big Bang to the formation of the Planet Earth. It has hit some bumps thereafter, particularly with any precise explanation of the beginning of life.

Modern science has encountered major difficulties in explaining, let alone reenacting in a lab, the beginning of life. Much research has been devoted to the question of whether organic life happened spontaneously a little less than four billion years ago. In my book, *On the Likely Origin of Species,* I describe the famous Miller-Urey experiments in the 1950's, which had portentous implications, and subsequent, less successful results in the efforts to replicate biogenesis. Suffice it to say that the effort to create organic life artificially has floundered. (Another book to read on this topic is *Undeniable*, by Douglas Axe.)

Explaining or replicating biogenesis is difficult enough. Much more difficult to explain is how the entire biosphere came together; it is a classic case of what came first, the chicken or the egg? Except there are myriad chickens and myriad eggs that had to appear just about simultaneously for the Planet Earth to function. (A book that explains the remarkable interdependence of all forms of life on this planet is *What is Life?* by Lynn Margulis and Dorion Sagan.)

Although biogenesis remains a mystery, and the origin of human life itself even more of a mystery, there is some consensus on the notion that our species is flawed in some way. In stark contrast to animals, which seem to inhabit the earth in a carefully calibrated clash of appetites, humans seem intent on disturbing what might otherwise be a wonderful, natural equilibrium. Unlike animals, humans seem quite capable of disobeying their nature and doing some irreversible damage to themselves and their heirs in the process.

Let's review how humanity's brightest minds have analyzed that tendency, as well as its cause.

Flawed Man

Ancient philosophers, as well as modern savants, in particular philosophers and psychologists, have struggled with what appears to be a defect in the human condition. Why do humans who have more than they need fail to share with those who don't have the necessities of life? Why do some prefer to steal the work of their neighbors rather than work for themselves? Why do we exaggerate our own accomplishments, while failing to take note of the insights and contributions of others? Why do the strong take advantage of the weak, the men take advantage of the women, the adults take advantage of innocent children?

Why, asked social scientists, do we humans have a tendency to want to destroy this planet?

For both believers and unbelievers, the biggest concern – given that God, if he/she exists, clearly allows bad things to happen – has been how to stem the flow of bad things. How exactly can our species change our *modus operandi* to prevent cruelty, conquer disease and stop wars before they end our collective existence?

In summary, those who want to believe in a benevolent creator, but are confronted with a very imperfect creation, ask ourselves: *Why do bad things happen to good people?*

The "God Theory," as presented in Genesis and other theological tracts that claim to have some divine inspiration, presents a viable explanation of what is clearly a flawed species, living in a flawed world. But it is not the only theory, and we should explore at least some alternative explanations for why good and evil seem to compete for ultimate control of our little planet.

Dualism as an Alternative Theory

There is one important alternative theory to the biblical notion of some sort of ancestral misconduct that led to an inherited trait that makes humans occasionally listen to our lesser angels. It is called "dualism" – which means basically that nature is half good and half bad.

Here's how Yuval Noah Harari explains it:

> *For dualists it's easy to explain evil. Bad things happen even to good people because the world is not governed single-handedly by a good God. There is an independent evil power loose in the world. The evil power does bad things.*

Harari admits that dualism has some major logical flaws, including what he calls the Problem of Order. Here's the quandary:

> *Dualism has its own drawbacks. While solving the Problem of Evil, it is unnerved by the Problem of Order. If the world was created by a single God, it's clear why it is such an orderly place, where everything obeys the same laws. But if Good and Evil battle for control of the world, who enforces the laws governing this cosmic war? Two rival states can fight one another because both obey the same laws of physics.... When Good and Evil fight, what common laws do they obey and who decreed those laws?*

The reader may find other theories to explain why, despite our individual tendency to stray from the path of goodness, there is today, more than ever before, an abundance of goodness and welfare in the world.

For me, there is only one viable theory – one coherent belief system that manages to connect what we know of science with what we intuit to be beyond empirical, testable science. Only one synthesis has been formulated that explains not only the "how" of things, but also the "why."

This is admittedly a personal opinion. I hope I can back it up in the rest of this book, using all the tools of both empirical and social sciences.

Hopefully, the reader will conclude with me that history's champion system of values is none other than the edifice of reason whose foundation was laid by Aristotle and whose upper floors were laid by Augustine and Aquinas. Perhaps not coincidentally, the last couple of centuries have witnessed a flourishing of that edifice, and added some upper floors.

These upper floors are more than theoretical constructs. They are actual societies built on the principles enunciated by the great savants and fine-tuned by the modern sciences. Those societies will be reviewed in later chapters. The reader can judge if they pass muster.

Getting back to the foundation, we find that it consists of the tenets of what is called the natural law. Among lawyers and philosophers, referring to the natural law is something that is seen as a high-brow idea.

It is not. One of the pithiest definitions was given by Erich Fromm. the famous German author of *The Art of Loving,* when he said that it consists of "learning to like to do what we know we need to do."

Let's dissect its tenets a little bit.

The Natural Law

To Thomas Aquinas and to most theoretical jurists over the last two thousand years, there is a prescribed way for humans to interact with nature and with one another. Humans come to know the natural law by instinct, by reason and by experience.

Not all thinkers agree that there is such a thing as a "natural" way of doing things, as opposed to an "unnatural way." In fact, the number of scientists who think that way has been decreasing; more in vogue now in academia is the view of scientific materialists, who don't buy at all into any notion of a natural law. Evolutionary biology, when applied not just to animals but to humans, has eroded the classical theory of the scholastics, which prescribes a normative code of conduct for humans.

Here's the way Yuval Noah Harari explains it:

> *In truth our concepts 'natural and 'unnatural' are taken not from biology, but from Christian theology. The theological meaning of 'natural' is 'in accordance with the intentions of the God who created nature. Christian theologians argued that God created the human body intending each limb and organ to serve a particular purpose....But evolution has no purpose.*

Harari clearly overstates the argument. Evolution may not have any purpose, in the sense that higher forms of organic life evolved by accident. But that clearly ended with the arrival of humankind.

Almost all human activities have a purpose.

Law has a purpose, which is to guarantee fundamental justice – so that all citizens can enjoy "life, liberty and the pursuit of happiness."

Medicine has a purpose, which is to keep humans not only alive but as healthy as possible, a concept which involves figuring out what nature intended for the various parts of the anatomy.

Psychology has a purpose, which is to keep the mind-psyche from becoming obsessed, deranged, addicted or otherwise inclined to irresponsible or anti-social behavior.

Intuitively, the overwhelming majority of humans accept rather readily the notion that humans not only have a purpose in life, but are actually, in the main, obsessed with trying to ascertain their particular role in the drama of their existence.

On the other hand, if the natural law is so ingrained, so logical, so common-sense, why is it that humans have a tendency to not want to follow it? And if that natural law was instilled into humans by a creator or Big Banger, why is it so hard to discern?

Three religious traditions, followed by more than two billion members of our species, have a rather intriguing theory to explain what appears as a disconnect between our species and the Big Banger. Jews, Christians and Muslims are convinced that in the ideal, original state of nature, it was easy for humans to discern what was natural and what was not, by the existence of a very clear channel of communication with themselves and their creator. Our earliest ancestors, they believe, lived in a paradise where instead of a "tangle," or tension, between God's will and human appetites, there was perfect coherence.

Something went wrong and the result was a disjointed, disoriented human family. Whereas initially humans spoke the same language as God and each other, the ancestral catastrophe resulted in a tower of Babel.

Henceforth, man fought with God and with other men – often over the silliest of things.

Let's review the theory presented by that model.

The Biblical "God Theory"

Certain of our ancestors, which people have designated as prophets, say it all started with a moment and place in time, called Paradise, when human forebears allegedly broke ranks with God, disobeyed his/her clear commands, and became tainted by what later became known as "original sin." From that moment forward, particularly in the Christian

Science and the Theory of God

tradition, the spiritual history of our species has been concerned with a reconnection between God and *homo sapiens*. That reconnection or reunion is supposed to bring coherence once again to our species.

In this chapter, we delve into that alleged, admittedly unverifiable history. We also apply the most elementary test of philosophy – which is simple logic, or what the layman refers to as "common sense."

We begin by looking at what we know of our earliest ancestors, in an effort to determine whether our species came into the earth already prone to engage in conduct that we often refer to as leading to "self-inflicted wounds." In other words, we must determine whether human nature always had a flaw, whether even the first humans suffered from the harmful addictions and impulsive tendencies that so often cause us to wish we could take something back.

After all, if humans were created by an infinitely wise and infinitely good Big Banger, why would we be flawed from birth?

Other than biblical stories and other legends and narratives, we don't have a lot of information about our earliest ancestors – beyond pictures that are particularly abundant in European caves. We will say more about those later. We will also review the annals of recorded history, dating back to Babylonian times.

What we do know is that for the last two millennia, the Western world has placed an astounding reliance on a story that has no particular scientific basis, but an amazing amount of sociological and psychological relevance.

It is the Genesis story, and it has shaped how the Western world thinks of our early ancestors.

Humankind's First Fall, According to the Book of Genesis

The idea that our species is flawed and that the flaw was somehow passed down from generation to generation first arose in the Book of Genesis, which is the very first in the Hebrew bible. Our alleged original ancestors were a couple reportedly named "Adam" and "Eve." (The names are meant to convey something about the origins of the first couple; "Adam" means "from the red earth" and "Eve" means "breath of life." It makes me wonder if the biblical authors were echoing modern, colloquial characterizations of men as worms that crawl on the ground or pigs that wallow in the mud, while women are angelic and sublime....)

Adam and Eve reportedly disobeyed God; as a consequence, they lost their innocence and their entitlement to residence in a legendary paradise where work was fun, pain and suffering absent, and even childbearing was as pleasurable as conceiving the child.

Adam and Eve were thrown out of Paradise. From that point forward, everything that our ancestors did was fraught with the kind of tragic flaw that characterizes Greek tragedies. We all know the trait, because we all share it. We see it in our kids, even at infancy.

Educators know that kids have to be taught to share, to listen to adults when they say, "don't get too close to the edge of the pool or you'll fall in" or "don't eat that fruit until it's ripe or you'll get indigestion." As they reach puberty, kids have to be taught to either avoid sex (the conservative advice) or at least engage in "safe sex."

The bad apples in our midst end us as juvenile delinquents. Some of those end up as recidivist lawbreakers. Others cheat on their taxes. Still others drink too much and get charged with driving under the influence.

A lot of those end up in jail. But even the good, solid citizens know, in our heart of hearts, that temptation lurks behind every sexual situation, every business deal that could be milked by a little bit of deceit, every confrontation with another member of our species who seems to be taking an unfair share of the highway or the marketplace. Even when dealing with those we love dearly, there is a tendency to exaggerate our own needs and ignore those of our spouse, or child or friend.

All of us know that there is a lure to being "bad." It is a universally shared personality trait in our species. Whether it's part of our nature, or an acquired trait, we know it's there.

And yet if that is the case, how can we explain the anecdotal evidence that shows so many members of our species that seem to have been born inclined to act responsibly and lovingly in all circumstances? Here I refer to those kind, motherly souls one often sees going about their lives doing good. (The reason I don't say "fatherly" is that I don't know too many totally blameless males.)

These kind souls don't seem tempted. Perhaps they lose their temper occasionally and say an unkind word; but in those cases, you can't help but think it's deserved by the object of their ire.

We have all read about couples who have adopted thirty and forty kids, providing them with a home that is almost the equivalent of what their natural parents would wish they had provided.

Selflessness is not necessarily a sufficient and necessary consequence of faith or religious inspiration. In many people, it seems to come naturally.

Celebrated author Karl Stern paints an interesting portrait of such natural virtue as it occurs so often in domestic workers of his native, pre-war Germany. In his memoir, Sterns lavishes admiration on women like Kati Huber, his wife's maid, who exude "the odor of hard work, the righteousness of the Psalms and the peace of the Gospel." He then characterizes their natural charity, which he calls a "treasure of anonymous sanctity."

Which is a very appropriate term for so many women whom I have met. There are countless wives and mothers and "aunties" and grandmas in the poorest neighborhoods of my city and county who display such "anonymous sanctity." The quality seems to be more in vogue among women – at least in my experience.

One, who is my niece, is worth describing in some detail.

Secular Ascetics: The Virtuous Aunt

I have particular admiration for my niece, who is fifty-years old and living a single life. She wasn't always a celibate and she doesn't now have a pledge of celibacy; by practice, if not by pledge, she is unmarried, romantically unattached, and rather fulfilled in her single life.

She lives in California, a continent away from my Miami home. For many years, we have collaborated on family matters; and I have found her very mature and empathetic insights enormously helpful in dealing with a variety of family issues – as are bound to occur in a family of fourteen siblings, fifty-some cousins and umpteen grand-nephews and grand-nieces.

My niece has recently been named pastoral associate in a church controlled by men. She recently showed true heroism when she came to the aid of a cousin who is a single mother with two small children and a dysfunctional male partner. During judicial proceedings brought by the state, the judge looked around and found one person, out of the entire gang of family litigants and observers, who was clearly fit to care for two children, ages two and four.

The judge chose my niece to be guardian of two little children who were not her own. I found out later that she had to ask her colleagues at work to cover for her during the four months of guardianship. She was working full-time and was not independently wealthy. She had to change her life, cut personal expenses, eliminate all leisure time, forget dating, forget a fun evening on the town, forget a siesta while watching her favorite movie or any other simple pleasure.

It was a case of instant family. It showed the kind of generosity of spirit that does not bear rational analysis. Her life had been changed by her generosity, her sense of Christian love, and a simple circumstance of fate. She was the only family member living in Los Angeles who could have come to the help of a cousin and her two kids. If she didn't accept the court's appointment, a perfect stranger would be chosen, with all the potential psychological damage to the children that such guardianships imply.

My niece says she was inspired by me – yet I never took guardianship of two little kids while living alone on a modest income.

In his marvelous book *The Great Divorce*, C. S. Lewis describes a scene where a soul in purgatory sees a spectacular woman approaching, and wonders if she is Joan of Arc or Mary, the mother of Jesus. Here's the gripping exchange from Lewis:

> *'Is it?....is it?' I whispered to my guide. 'Not at all,' said he. 'It's someone ye'll never have heard of. Her name on earth was Sarah Smith and she lived at Golders Green.'*
>
> *'She seems to be...well, a person of particular importance?' 'Aye. She is one of the great ones. Ye have heard that*

fame in this country and fame on Earth are two quite different things.'

Lewis was perhaps trying to tell us something that all the great theologians have missed: that both sin and redemption cannot be defined by a one-size-fits-all equation. Perhaps the tendency to be selfish, arrogant, petty, and occasionally cruel does not affect all of us equally. *Perhaps not all of us need major psycho-spiritual rehab.*

One-Size-Fits-All Original Sin

From the scientific standpoint, it is increasingly plausible that all humans have an inherited flaw. But there's nothing to say that it affects all of us to the same extent.

Everything we know, and just about everything we read is consistent with the notion that our earliest ancestors were created by the same infinite being that propelled the Big Bang. If that is true, then it would be logical that they possessed an undiluted innocence, a pure inclination to do good and be good.

The very idea that humans were originally blameless is wholly consistent with the idea of a benevolent God. And so is the idea that in that original, primeval state of unblemished existence, they could view more clearly what was good for them to eat or do or say.

Assuming that, it is entirely plausible that they disobeyed their nature and ate or drank too much, or plucked the fruit from the wrong tree. Perhaps they "lost their innocence" by failing to fulfill a love vow or by stealing the fruits of another person's hard work.

If so, this improper behavior, this clear and conscious deviation from what we have called the natural law caused things to go awry. That a defect in the DNA was transmitted, either in a Lamarckian way or in a Darwinian way that we don't fully grasp, to their offspring. And that it remained for many generations, until the estranged, formerly creator-friend and now distant-in-time Big Banger took it upon himself/herself to remedy things.

That first, ancestral sin was later accepted into Christian doctrine as "original sin." It was a one-size-fits-all burden that weighed on the most talented and least talented, the richest and the poorest, the physically powerful and the physically or mentally challenged.

In one celebrated case, the concept was fine-tuned by a brilliant and immensely flawed man named Augustine of Hippo. He was later known as St. Augustine, and he refined the concept of original sin by his introspective writings of the fourth century after the birth of Jesus.

Augustine: From Bad Boy to Saintly Philosopher

Arguably one of history's three greatest philosophers (along with Aristotle and Aquinas), Augustine played a decisive role in developing the church's understanding of original sin through his autobiographical book "Confessions."

Augustine's ideas were shaped to a great extent by personal experiences – most notably his inability to forego frenetic sexual adventures. The juxtaposition of his Christian beliefs (acquired later in life) with his sexual proclivities (the staple of his youth) provided a valuable intuitive laboratory for the analysis of free will.

Intellectually, Augustine had absorbed a value system that mandated strict observance of monogamy; biologically, his appetites drove him in the opposite direction. One was an accelerator; the other a brake. In his view one was the result of sin; the other a moral imperative that reflected a well-ordered creation by an all-knowing God.

The theory, as previously explained, begins with the notion that our species needs to be "redeemed" – that it needs to be transformed from a lesser form of existence to a higher form, like a caterpillar that needs to morph into a butterfly.

Augustine begins the analysis by admitting his addiction, saying: "For my will was perverse and lust had grown from it, and when I gave in to lust habit was born, and when I did not resist the habit it became a necessity." He is surprised that his newfound desire to follow what he perceives as God's law is blunted by his acquired habits, which seem to have a will of their own. Augustine elaborates:

> *"So these two wills within me, one old one new, one the servant of the flesh, the other of the spirit, were in conflict and between them they tore my soul apart."*

He proceeds to define the condition as a form of mental illness, a pathology which he calls a "disease of the mind." This idea resonates and meshes quite well with many modern social scientists, particularly those who study addiction (to drugs, alcohol, sex) and the treatment thereof.

Psychologist and popular author, Arthur Koestler (author of *The Ghost in the Machine*), describes the apparent flaw in our species as a "streak of insanity which runs through the history of our species and which indicates that somewhere along the line of its ascent to prominence, something has gone wrong." He follows that characterization by arguing that it does not conflict with evolutionary ideas of survival of the fittest, saying:

> *Evolution has been compared to a labyrinth of blind alleys; and there is nothing very strange or improbable in the assumption that man's native equipment, though superior to that of any other living species, nevertheless contains some built-in error or deficiency which predisposes him towards self-destruction."*

Whether it's a "streak of insanity" or a "built-in deficiency," it's evident that modern scientific ideas about the flaw in our species are actually quite similar to those found in the biblical narrative – minus the reference to "sin."

Koestler reaches that conclusion, stating it in rather blunt terms:

> *The ancient doctrine of original sin, variants of which occur independently in the mythologies of diverse cultures, could be a reflection of man's awareness of his own inadequacy, of the intuitive hunch that somewhere along the line of his ascent something has gone wrong.*

Other modern scholars, as we shall see, prefer the term "vulnerability" to refer to this perceived built-in defect or inadequacy. (I have to admit I like Koestler's more colorful characterization of the flaw in our species as a "streak of insanity.").

But let's see how modern social scientists explain a trait that no one seems able to deny.

Original Sin = Original Vulnerability

Scholar and associate pastor Ellie Hidalgo, in a recent (2010) graduate symposium at Loyola-Marymount University, first explains and then modernizes St. Augustine's intellectual journey, leading to what is, in the Judeo-Christian tradition, the generally accepted doctrine of original sin:

> *Seeking a reasonable explanation for why humans are so incapacitated, Augustine references Scripture, namely St. Paul's Letter to the Romans 7:17 in writing, 'My action did not come from me, but from the sinful principle that dwells in me.'*

Ms. Hidalgo's analysis continues by delving into the biblical roots of the concept:

> *Augustine then reaches even further back to the beginnings of the Hebrew Scripture to explain why humans carry a sinful principle and lays the blame at Adam's feet. It is the belief in this sinful principle, originally created by Adam's fall and passed on from generation to generation, that comes to be known as the theology of original sin.*

Hidalgo concludes the analysis by using modern scientific terms to describe what Augustine implies:

> *The implication is that from birth a child contends with being in a state of inherited, ontological corruption. Humanity inherits Adam's original propensity to abandon good and choose evil.*

Citing modern authors, she calls that propensity "original vulnerability," instead of the traditional "original sin."

As we shall see, it is as good an explanation of what appears as a human tendency to stray from the ideal path – from what the Greeks called the "virtuous" path and into the realm of what modern sociologists and psychologists refer to as "anti-social behavior."

Let's accept that premise for a moment. Let's assume that humans are flawed beyond spontaneous self-repair, of the kind that happens constantly in almost every part of our bodies. It is not too far-fetched to believe that human DNA is lacking in some essential ingredient, perhaps as a result of some contamination that took place when our ancestors ate the wrong food, or ate the right food, but in excess portions.

In the next chapter, I touch upon a new theory which actually suggests there could be a genetic passing of mutations acquired by one generation to the next. We will get to that, as well as other scientific indications that certain pathologies could be the result of repeated, willful actions that are passed down from generation to generation.

Let's also discuss how to cure or lessen what is apparently an inborn, pathological condition. Is there an antidote or treatment mode we can administer to remedy the situation? Or do we just need to bear down, accept our imperfect lot, and wait for the reward of an afterlife, as most Christian denominations believe?

My dad framed the solution in terms of a very simple equation. To overcome our human tendency to placate our appetites, one needed self-discipline. The incentive to apply self-discipline came from the sheer math of what he called "the economy of salvation." You weigh a temporary pleasure (drinking, sex, money, power) against an eternity of joy with the creator and the solution was clear: it was called "delayed gratification."

Put simply, it means that eternal happiness trumps seventy or eighty years of gluttony and intemperance. As historian Chris Wickham explains, the presumption was that –

> ...this world was only a brief testing ground before the eternal joys of heaven or the eternal tortures of hell, which meant

> *that pleasure was risky, and that asceticism, sometimes self-mortification, was increasingly seen as virtuous.*

The same prescription, applied to society as a whole, worked rather well to maintain order and a measure of welfare for all. And, as Wickham explains, no one was left behind, because "these beliefs went generally with an exaltation of poverty – for the good Christian ought to give everything to the poor…"

Eventually, the frugal quality of the Christian prescription led to a culture that valued thriftiness and hard work. When that ethos was exported by the British to North America, it came to be identified with a reform strain of Christianity that went by the name of Protestantism.

The Protestant Work Ethic

The German, Dutch, French and English-speaking Europeans who settled the northern regions of North America in the seventeenth century were mostly Protestants. These reform-minded Christians thrived on discipline, hard work and a sense that God watched closely over the human flock with a view to castigating the unrepentant, sinful ones and rewarding the virtuous ones with "joys that never end."

Their culture saturated the various regions of what we now call the United States, except for Florida and Louisiana, where Spanish and French Catholics initially ruled the roost.

Like my father's notion of an "economy of salvation," their recipe was what is referred to as the "Protestant work ethic." If we stuck by such ethical norms, we would earn what Christians call "beatific vision" (meaning an eternity in the company of the Big Banger).

For me growing up, the idea of adding good deeds and subtracting bad deeds to determine one's eternal fate sounded pretty straightforward and logical. It made enormous sense both individually and collectively: If the great majority of people followed the Protestant work ethic, you would have happy, fulfilled, educated individuals. Moreover, this would lead to a society that functioned smoothly under common values that generally did not have to be enforced by the civil authorities.

Political scientists and sociologists even used quantitative methods to show that having an internal moral compass made sense for society as a whole. If one assumes no internal moral code, they argued, the law would have to be enforced by a near-infinite number of police officers, watching what every individual did and punishing the transgressions immediately, before anarchy and mayhem ensued.

It was easy to conclude that a common moral code would blunt what Koestler called our human "streak of insanity." To me, it was all self-evident. People doing the right thing (e.g., the Ten Commandments) enjoyed healthy lives, with lasting marriages, thriving economies and peace based on justice and respect for all other humans, regardless of race, color or religion.

Afterwards, we would go to heaven, which was a place that my dad defined as "all good without any mixture of bad." Hell, by contrast, was a place where we could expect what my dad defined as "all bad, without any mixture of good."

As I delved into modern sciences like genetics and psychology, the entire matter of salvation – of heaven and hell – took on a more nuanced character.

I looked for inspiration in the writings of history's great poets, hoping for some subtlety and romance. The Protestant ethic and the Ten Commandments constituted a rational enough foundation in law; but I wanted something that appealed to the heart as well as the mind. I turned to the most acclaimed writer who sought to describe heaven, hell and purgatory.

I turned to Dante.

Dante's Circles of Hell and Heaven

Dante ranked the circles of hell in the kind of quantitative sequence that would have made my dad happy. But it was just an author's technique. (It reminds me of rabbis who summarized Moses's commandments into ten rules, for the simple reason that we have ten fingers....)

I was glad to find in Dante a very refined concept of the transition that a human has to endure in order to join the creator. As we shall see, it is not too different from what modern writers (e.g., C.S. Lewis) and modern psychiatrists (e.g., Scott Peck, M.D.) envision as a transitional place between the human condition and a place or state of mind in which dead humans get to spend time with our creator.

Here's how Dante describes the process of redemption in a transitional place which Catholics call "purgatory":

> *You have escaped damnation and made it to Purgatory, a place where the dew of repentance washes off the stain of sin and girds the spirit with humility. Through contrition, confession, and satisfaction by works of righteousness, you must make your way up the mountain. As the sins are cleansed from your soul, you will be illuminated by the Sun of Divine Grace, and you will join other souls, smiling and happy, upon the summit of this mountain. Before long you will know the joys of Paradise as you ascend to the ethereal.*

The theme of purgatory, or a place where one must go to "purge" oneself of certain baggage that has been acquired over a lifetime, also fits in nicely with modern psychology. Modern psychologists universally advise us to discard our "hang-ups" or our "baggage," meaning those things that detract from our ability to share, relate and collaborate with others. Peck uses the same metaphor, describing our journey from earthly life to heavenly life as a journey in which it behooves us to "travel light."

That concept rings very true to me. There is so much of us that is artificial, plastic, needless, self-defeating. None of that baggage defines the healthy members of our species. None of it is an essential part of what makes us human. None of it makes us grow or love or feel fulfilled.

It seems evident that a clean heart (scientifically, a "mind-psyche") is part and parcel of a happy human. And happiness requires more than just physical welfare; it requires a harmonious combination of the physical with the spiritual.

Dante, once again, provides a more lyrical description of the phenomenon:

Science and the Theory of God

> *The varying voices of the spirits in the different Heavens blend into a sweet harmony. 'As varying voices make sweet harmony on earth, so is it this holier sphere. Degrees of difference make one song entire; and various flames construct one wheel of love.'*

Not all religions predict an eternal afterlife. But the ones that do suggest that it consists of a permanent state of happiness, defined by whatever in the human experience constitutes happiness. We will discuss that issue in various sections of this book, bringing to bear modern insights that flow from both empirical and social sciences.

We will also discuss to what extent there is some quantitative fairness to it, in terms of evening the score for those who are short-changed by earthly life. In other words, there is obvious merit to the equation that argues for some measure of delayed gratification. It doesn't take a genius to figure out that eternity, if it exists, is a much longer period than earthly life and can compensate for shortcomings endured while alive.

But is eternity a logical or necessary reality? Does an afterlife explain or complete our earthly life to such an extent that without it, our lives are meaningless? Or is the religious belief in heaven as the "opium of the people," the escape that Karl Marx proclaimed it to be – the stratagem used by those in power to manipulate the masses into accepting a lesser lot in this life?

Our analysis begins with the simple fact that humans, alone among all the species that crawl on this earth, can't accept a limited lifespan.

Why Is Heaven Necessary? Why an Afterlife?

Unlike the animals, who seem to accept death rather docilely, humans resist death with all their might. Humans don't want to die. Americans, for example, spend more than a third of all health-care costs in the last six months of life.

Think about that. Americans have life expectancies in the range of 75 years (combining men and women). Yet they spend more than a third of their resources devoted to health care on the last 0.6% of their time on this earth! And, I should add, the amount they spend on health care is enormous: fully one-sixth of the entire gross national product, equivalent to close to three trillion dollars.

One-third of three trillion dollars is one trillion dollars, which is half the total gross national domestic product of Brazil. Americans spend that much on trying to keep themselves alive for the last six months on this earth. They clearly want to prolong life as much as possible.

Humans are traumatized by the realization of a short life span and conversely enthralled by the possibility of eternity. Here's how the great scientist, Louis Pasteur describes the fascination:

> *I see everywhere in the world the inevitable expression of the concept of infinity....So long as the infinite weighs on the human mind, so long will temples be raised to the cult of the infinite, whether it be called Brahman, Allah, Jehovah or Jesus....The ideals of art, of science, are lighted by reflection from the infinite.*

And the yearning for infinity goes beyond rhapsodizing about it in song or conceptualizing it through science. Humans aspire to eternity by wanting to be an integral part of it. As the saying goes, "nobody wants to die, but everybody wants to go to heaven."

The world's religions haven't always agreed on the details of how exactly our species earns an eternity of happiness in the company of the presumed almighty creator. Even among followers of Jesus, there have been intellectual struggles on the way to a heaven or paradise.

For a while, it wasn't clear whether what counted was prayer or good works.

Prayer v. Good Works

At one point in the history of Christianity, there developed two rather distinct theories on the issue of how humans could be reconnected to the creator. One said that what counts is prayer, since it is only by divine power than we can possibly reach heaven from our lowly, prostrate, human condition. The other said that what counts is what we do: that we must *earn* heaven.

Modern theologians of all Christian denominations pretty much agree that both prayer and good works are important. *A consensus has been reached that prayer is the way we discern what God's will is, and that doing God's will is what merits eternal life with God.*

As we shall see, this holistic view of what it takes to earn a place close to the Big Banger for all eternity seems quite compatible with science.

But even within the Catholic Church, there is a clear divergence in tone. One side, which we can call the "Valley of Tears" faction, preaches that we should be tearfully repentant, other-worldly rather than worldly, and not in any way contented with our lot as "worms" who crawl on the surface of an animalistic planet.

It is a rather dreary view of humankind, which turns off many people who would otherwise consider themselves adherents of Christianity.

The Valley of Tears Model

I could use many vignettes to illustrate the Valley of Tears model of human redemption. But the most recent one involves one of my sisters and the season of Lent. As previously mentioned, for Catholics this is a time of reflection and penance, when one is supposed to empathize with the sufferings reportedly endured by Jesus.

My sister is a pretty faithful observer of her Christian faith. But she just hates the whole idea of Lenten observance. This year, when we were right smack in the midst of Lent, she blurted out: "I hate Lent. It is totally inconceivable to me that we should invent sacrifices to add to the ones that life presses upon us. It makes no sense to me to invent misery or suffering when there is so much already there."

Her view is shared by what is probably a majority of Catholics – if not other Christian sects. In the old days, the preachers who based their homilies on that style were referred to as purveyors of "fire and brimstone."

That faction of Christianity is characterized by their emphasis on the fallen character of our species, rather than on the potential good that humans can do, if they attune their natures to a presumably benevolent god.

You can call them the pessimists – as opposed to the more optimistic approach of the modern church, represented by Pope Francis. That faction sees our physical existence as the beginning of a journey that, by all rights, should be fun-filled and compatible from its earthly beginnings to its heavenly end.

The Seamless Transition Model

Catholics of the Pope Francis school tend to connect the joys of heaven with a joyful life on this earth. They have mostly discarded the notion that earthly life is a painful stage that we must endure on our way to our true destiny in the company of God and his angels. They align more with the notion, espoused by St. Anthony of Padua, that in our earthly existence we construct, as it were, the dwelling which will be our abode for the balance of eternity.

I call this the "seamless model" of redemption – which is kind of the opposite end of the spectrum from the "valley of tears" model.

The notion that we build during this life the bed in which we lie in heaven makes enormous sense to me. If we assume an infinitely wise and benign creator (a plausible assumption for reasons previously stated), then it is logical that there should be a *seamless connection* between our allegedly temporary stay on earth and our hoped-for, potentially unlimited stay in heaven.

The seamless model of redemption is also more compatible with the insights of modern psychology, which finds affinity between theological concepts of what is desirable human conduct and what is healthy behavior by our species. Because there must be harmony between what science dictates and what theology mandates, the person who follows what we have previously referred to as the natural law will tend to have better health in this life and a better chance for eternal life.

Besides being more compatible with science, the seamless model of redemption is more humane and alluring. Perhaps for that reason, it is the one clearly espoused by the current pope.

The Joyful Christian

Nothing and no one defines the modern prescription of a virtuous, and also joyful, life like Pope Francis. An Argentinian Jesuit who discards the trappings of ecclesiastical pomp and circumstance, Pope Francis has redefined Catholic ethics – with both his pronouncements and his example. In words and deeds, his vision of Christianity deemphasizes sin and emphasizes compassion and tolerance.

It's no wonder he declared 2015-16 as a celebration (jubilee) "Year of Mercy." What may have, in an earlier era, been some sort of a papal edict is instead a catalog of things that Jesus did when confronting four real people, each flawed in a visible, tangible way.

Let's list them:

1. Zacchaeous: His defect is that he was "enslaved by money."
2. The adulteress: Her defect was "seeking happiness only in created things."
3. Peter: He betrayed Jesus and was moved to "weep" when confronted by him.
4. The repentant thief: He was "assured Paradise" after a lifetime of bad conduct.

The four sinful characters listed above illustrate clearly the theological views of the man who leads the largest Christian denomination of all, with over a billion members. In the mentioned prayer, Pope Francis emphasizes (better than a scholarly encyclical could) the way of salvation for a modern Catholic: In all cases, it is encountering Jesus, asking forgiveness and amending one's life. One of the sinners loves money, another loves sex, a third lacks courage to do the right thing and a fourth is the classic, recidivist criminal.

As we shall see, modern psychology proposes a set of norms that define a mature person, that overcome or endure afflictions, and that enable, as much as genetic and environmental deficiencies will allow, the full flowering of the human person.

To be a truthful path back to reconnection with God, any religion, of necessity, has to be holistic. That is how the great Lutheran theologian, Dietrich Bonhoeffer, understood the combination of healthy traits that define the joyful Christian. One of his most eloquent biographers, Eric Metaxas, elaborates:

> *This was how Bonhoeffer saw what he was doing. He had theologically redefined the Christian life as something active, not reactive....It was not a cramped, compromised, circumspect life, but a life lived in a kind of joyful, full-throated freedom....*

As we will see in subsequent chapters, modern psychology lends credence to the Christian virtues as a healthier, more mature way of life.

But before we get to that, we need to tackle an anomaly that, for deeply flawed humans like me, is both a pleasant and an unexplainable surprise. I refer to people who seem to be "naturally" (by which I mean effortlessly) good.

Altruism and the Common Man/Woman

I once heard a preacher remark that the virtuous, faith-based life formula is a win-win proposition. Even if it turns out that there is no eternity of heavenly bliss for the good Christian (or the good Jew or Muslim or Buddhist), the observance of the Ten Commandments is a recipe for happiness here on earth. In other words, even if there is no Christian heaven, living a good, Christian life means a heaven-on-earth.

There are many holes in the argument, of course, including the proverbial "bad things that (often accidentally) happen to good people" and the intentional harm suffered by the good person who doesn't protect himself from the not-so-good evildoers that cohabit this earth with him.

Theologians argue that the key is to choose the ultimate good, not the proximate good. They admit that doing good often has bad consequences for the do-gooder; often, if you turn the other cheek once you've been slapped, you get slapped even harder. But, they argue, in the end, the equation favors the good person, since all negative accounts are settled in heaven, where the credits are, by definition, infinite and thus result in a positive balance.

So the equation is occasionally more like "short-term lose and long-term win." In the case of celibate priests and other religious ascetics, their entire life equation is premised on serious up-front losses and delayed gratification.

Like the virtuous aunt, the celibate monk and the unmarried nun seem to be in a totally different world of perfectibility in this life. It is not that they are immune to temptation; quite the contrary. It is more like an unfathomable, irrational, voluntary abstention from some of the fair gifts of humanity.

Are they blinded by faith? Are they so absorbed in the "valley of tears" model that they follow a "vocation" (calling) from some otherworldly source as a way to teach other members of the species how strong they feel about an eternity of love?

The whole history of Christian missionaries is an enigma to me. And, to the modern, materialist value system that permeates media and academia, it is (to borrow from Churchill) a riddle wrapped inside an enigma. Until recently, I was not familiar with the Middle East chapter of that history.

Modern Missionaries in the Middle East

Michael Oren is a former Israeli Ambassador to the United States and a well-published historian. I never met the man, and had never read any of his books, when something really odd happened.

One day, as I was returning to my car after a business meeting, I found a brand new book propped up against the door of my little roadster. The book had no dedication, markings or notes of any kind. It was ostensibly an anonymous gift.

My car was almost unique in Miami. It was a 1999 BMW M-roadster; its color was particularly attractive, to the point that even little kids were enthralled when they saw me drive by. (It was listed as "Estorilbleau" by the German automaker.)

I have to believe the person who placed the book next to my door knew I was the owner. Whether that is the case or not, I certainly felt compelled to read it.

Its title was "Power, Faith and Fantasy."

Oren's book begins with the story of Christian missionaries who risked their lives and gave up their wellbeing to spread the Christian message in lands long controlled by regimes which did not profess religious freedom. Like others who traveled to Africa, Asia and the Americas, the Middle East missionaries were often martyred.

They are examples of altruism that belies the flawed character of humankind. It is a phenomenon hard to explain – except perhaps if one assumes the "valley of tears" model. The altruist accepts short-term denial and sacrifice in order to earn a happy eternity.

But it is one hell of a gamble.

A more benevolent, satisfying and holistic view is the one that sees this life as a continuum to an after-life in which every human trait, every pleasure, every brilliant insight into truth and every loving relationship that fills us with earthly joy is increased towards a never-ending high. I have previously referred to that as the "seamless transition" model of redemption.

Later on, we will explore what modern science says about that model and how it might be harmonious with science. Here I want to share insights derived by an author who managed to portray the biblical story as mythology.

His name was C.S. Lewis.

C.S. Lewis and Purgatory

The best analogy I have ever heard or read about the passage to heaven is the one contained in a little book by C.S. Lewis, entitled "The Great Divorce." (The title, as Lewis explains, has nothing to do with marital break-ups; instead, it is taken from Dante's literary effort to "marry" heaven and hell. Lewis felt that heaven and hell ought to be "divorced" instead.)

But I digress. What makes C.S. Lewis's work enthralling is the mythical yet plausible description of the passage from death to eternal bliss, defined by being in the company of the Big Banger. Lewis, who was of the Anglican faith, portrayed purgatory as a voyage in which souls interacted with more established residents of paradise, who counselled them and encouraged them to continue their travels towards the infinite being waiting for them.

One of the more picturesque metaphors he used was the pain endured by souls in purgatory from being pierced by the blades of grass on which their bare feet trod. It is clearly Lewis's way of depicting the necessary pain that comes when humans divest themselves of excessive self-love.

Lewis's mythology of purgatory marks the final chapter of theological wars that have raged over the last twenty centuries among Christian denominations. Thankfully, Christians no longer fight about the fine points of the Jesus story. They are nearly unanimous in their agreement that *what counts is a clean heart and a sense of service towards others, coupled with earnest effort to promote the common good of all.*

But only C.S. Lewis, of all Christian writers and theologians that I have read, has a theory that is able to reconcile short-term misery, including premature death, with any kind of divine plan. We have all heard the saying that children must expect to bury their parents, but it is plain unfair to have parents bury their children.

So how does the Big Banger correct the unfairness?

C. S. Lewis offers not only a qualitative but also a quantitative explanation of what we can call the divine "balancing of accounts," when he says:

> *That is what mortals misunderstand. They say of some temporal suffering, 'No future bliss can make up for it,' not knowing that Heaven, once attained, will work backwards and turn even that agony into a glory. And of some sinful pleasure they say 'Let me have but this and I'll take the consequences': little dreaming how damnation will spread back and back into their past and contaminate the pleasure of the sin. Both processes begin even before death. The good man's past begins to change so that his forgiven sins and remembered sorrows take on the quality of Heaven: the bad man's past already conforms to his badness and is filled only with dreariness. And that is why...the Blessed will say 'We have never lived anywhere except in Heaven;' and the Lost, 'We were always in Hell.' And both will speak truly.*

Sounds good, but it is still hard to conceive that a man (or woman, though I am not much of an expert on that) can willingly give up sex, having a family of his own, and living an ascetic life, while surrounded by equal or less talented people who enjoy the nice car, the expensive wine or the penthouse by the water.

I should add that celibacy totally confounds the scientific materialists. Of all the forms of altruism, voluntary celibacy seems the one that most contradicts their theories, which are based on the single instinct to survive and procreate. Here's how Yuval Harari puts it:

> *The Catholic alpha male abstains from sexual intercourse and raising a family, even though there is no genetic or ecological reason for him to do so.*

I have met many celibates, but shared an apartment with only one. His level of altruism, coupled with his charismatic voice, his multilingual skills, his sense of humor and his manly, good looks combine to make him a great candidate for sexual interaction and accumulation of riches and other creature comforts. Like many other celibates I have met, he could have been quite successful in the secular world.

Yet he chose a life of personal poverty and abstinence.

Considered together, the naturally virtuous aunt and the celibate priest beg for explanations that are hard to find in science alone. The same is true of the "bad-boy-conversion" story of St. Augustine. Without positing the existence of a God and an after-life, I am, frankly, unable to explain my own struggles with ordinary temptations like excessive alcohol or abuse of power.

Observing and pondering the different paths to greater goodness and less selfishness does serve to illustrate a principle that will be developed throughout this book, which says that humans are not all created theologically equal. The distribution of traits and talents is by no means even as it worked its way from the first humans to us.

Some of us fit the description that Churchill gave of one of his adversaries, when he said the man "was modest, and had much to be modest about."

Others seem unable to abide by even the most elementary rules of society. I have served in city and county government with quite a few who, having a decorous lifestyle, could not refrain from wanting more. One, who served with me on the city commission and later was indicted on bribery charges, was so ethically challenged that the famous former attorney general, Janet Reno, said of him that "he had larceny in his heart."

Another, who was brilliant and quite powerful as a county commission chairman in a county of almost three million people, ultimately succumbed to depression and committed suicide in the foyer of Miami's daily newspaper.

We are not born with the same, exact amounts of avarice, or libido, or vanity in our hearts. And we're certainly not all capable – even if we will it with all our might – to lead heroic lives of celibacy and poverty.

I have been lucky to share an apartment with one who willingly and joyfully embraced the ascetic lifestyle from very early in his life.

The Pope's BFF (Best Friend Forever)

For this I have to go back to 1971, when I met this extraordinary man. I met him as I was trying to decide where to go to law school. "Padre Sean" (pronounced "Chan"), as his Hispanic parishioners in D.C. referred to him, was a young priest of the order of Capuchins, which are reformed Franciscans.

They are seriously into poverty – as in having literally no possessions except what is needed for survival in an American city.

Padre Sean gave me some very good advice, and I chose Harvard Law over Duke Law. Later on in life, I came to realize that his advice helped enormously to launch my career in politics. (Harvard, after all, is viewed by the media as the pinnacle of the desirable liberal education that all politicians should have, while Duke is despised by many in the Fourth Estate as a bastion of Southern, white elitism.)

Almost half a century has passed since I met him and spent a summer sharing a little, ratty apartment with him; Padre Sean O'Malley has gone from humble, sandal-wearing Capuchin friar to the person who Nora O'Donnell introduced in a recent "Sixty Minutes" piece as the Pope's confidant and "BFF" (best friend forever).

The question is: what would I have been without the inspiration of Cardinal Sean? What would my niece have done with her life without the example of an uncle who is a political leader? What would any of us be without the inspiration of a Joan of Arc or a Martin Luther King, an Abraham Lincoln, a JFK, or a Mother Teresa, or of the person who inspired all of those – a Jewish carpenter named Jesus?

And the related question: why should my potential eternity of joy or damnation depend on the luck of having good friends? Why should my after-life, assuming there is such a prospect, depend on who happened to be around to inspire me?

Much of this book is devoted to that question. But before we tackle that particular complication, it is necessary to establish, with some scientific certainty, that humans actually *choose* to be altruistic. That we do, indeed, have free will.

The ancients assumed that we did; to them, the person who acted in a benevolent way towards others (at least those within their social class) were "virtuous." They valued beauty, truth and peace over ugliness, deception and unprovoked violence.

And they assumed that our choices affected all of those outcomes.

From Socrates, Plato and Aristotle to Aquinas

The Christian concept of the human person evolved from the ancient Greeks. More specifically, as explained by Dr. Ferdinand Santos (in "The Social Anthropology of Catholic Thought"), the "triumvirate" of Socrates, Plato and Aristotle saw the human person as "inexorably directed towards truth." He contrasts that with other philosophers of the time (the Sophists), who linked truth "with appearance and opinion" and argued that the ability to manipulate this kind of 'truth' is what ultimately defines 'happiness.'"

No doubt my reader immediately grasps why lawyers are often accused of "sophistry" (being like the Sophists) in that we are often observed manipulating truth to fit what appears to a judge or jury as being favorable to their clients.

Christian philosophers, by contrast, including and most notably

Science and the Theory of God

Augustine and Aquinas, accepted the Aristotelian view of an objective truth that humans were bound to seek and put into practice.

Augustine and Aquinas also based their analysis of the human condition on the premise that in the original state of nature, the body and mind were aligned with the soul. They were certain that our ancestors were born quite able to act responsibly and lovingly.

Interestingly, I have anecdotally confirmed that a number of our species, even now, seem to have retained (or maybe acquired) that primeval innocence.

Let's take the example of those kind, motherly souls one often sees going about their lives doing good. These kind souls don't seem tempted. Perhaps they lose their temper occasionally and say an unkind word; but in those cases, you can't help but think it's deserved.

My wife's distant cousin is like that. She is not well off; and she has a house full of discarded relatives, pets, and strangers. They feast on her charitable nature. She never says "no."

It is entirely possible that Adam and Eve, or whoever were the earliest individual humans or tribe of humans, also possessed that innocent quality; that they could view clearly what was good to eat or do or say. It is entirely believable that they disobeyed their nature and drank or ate too much, or from the wrong tree. Perhaps they "lost their innocence" by failing to fulfill a love vow or by stealing the fruits of another person's hard work.

If so, this improper behavior, this clear and conscious deviation from the natural law, caused things to go awry. Biblical texts, and the tradition they spawned, suggest that a defect in the DNA was transmitted, either in a Lamarckian way or in a Darwinian way that we don't fully grasp, from our ancestors to their offspring. And that it remained for many generations, until the estranged, formerly creator-friend and now distant-in-time Big Banger took it upon himself/herself to remedy things.

But those are all unproven tales, mixed in with the intuitive reasoning of the greatest classical minds. Aristotle and Augustine were bright guys. Yet the most important trio of empirical sciences (physics, chemistry and biology) were embryonic until the Middle Ages. And atomic physics, genetics and microbiology were not developed until the last two hundred years. Toiling before that, and also before the sciences of psychology and neurology, a Dominican monk did his best to explain the mystery of the human mind-psyche.

His name was Thomas Aquinas.

Aquinas Describes the Human Mind

Aquinas had no laboratory to work from; he had little grasp of biology and none at all of psychology – which had not been invented yet.

He did most of his analysis by "deductive" reasoning, which was the premier analytical tool available to scientists of his time. Astronomers and mathematicians did have some grasp of what later became the elements of geometry, physics and chemistry, but even those empirical sciences were primitive in the thirteen century, when this incredible mind walked the earth.

Nevertheless, and by using an amazingly prescient form of introspective analysis, Aquinas was able to formulate what turns out to be a very precise description of the human mind. For us today, familiar as we are with computers, dividing the human mind into three functions (memory, understanding and will) would not seem far-fetched. The reason is that those are the three essential components of a modern computer, which relies on hardware full of *memory* bits, programs for analyzing (*understanding*) that memory, and the operating wizardry of someone *willfully* punching keys.

Clearly, Aquinas's troika of components of the human mind is very useful in a qualitative sense. Now let's try to quantify it a bit.

Scientists have calculated that a computer capable of processing the information that a human brain processes would be the size of a galaxy. Even with the most micro of micro-chips, we cannot replicate it.

The Human Mind as a Computer

But that still would not get us the human mind. Humans do calculations similar to a computer and post them in analytical papers, as a computer does. Admittedly, a computer is much faster and can do, for example, statistical regressions much quicker than humans. Regressions are used to determine variables that are correlated (such as smoking and lung cancer); they require many, iterative calculations. The same is true for analyses of storms and hurricanes, or of the path of a satellite.

After the computer analysis is completed, the human mind makes a decision based on the analysis. The decision may be to stop smoking so as not to end up with cancer; or to change eating habits; or to not contradict the boss to her face, even though our gut is bursting to rebel over some unfairness in the workplace.

What is amazing about humans is that each person makes a slightly different decision based on identical information received. We marvel that flocks of birds react instantaneously and identically to a wind that blows or a noise that is heard in the distance. The phenomenon is known as "synchronicity" and boggles the mind of physicists and mathematicians, for the simple reason that it seems to happen in unison. *There is no time for even the bird eyes to send a signal at the speed of light.* Yet these animals react together, simultaneously, instinctively.

It is clear that animals, like mega-computers, have pretty powerful hardware and pretty sophisticated software.

But they don't have the flexibility or creativity of the human mind. You can teach a computer to play chess, which is enormously complicated in terms of possible combination of moves. You can design a robot to do intricate movements across the ocean floor looking for shipwrecks and machines that burrow deep into the earth to mine precious metals or scan the heavens to predict weather patterns. But those sophisticated machines would not be able to tie a shoelace by observing a five-year-old human.

Or ride a bike.

Google, which is one of the world's most advanced companies, allocated a billion dollars to design a car that drives itself. It took me all of two hours to teach my teen-age kids how to drive a car.

Perhaps the greatest deficiency that computers have, in regard to humans, is their inability to communicate with other computers so as to mutually learn and increase their combined knowledge. The same is true of animals; generations upon generations of dogs chase cars fruitlessly and don't manage to teach their young not to repeat the experiment. Generations of cows allow themselves to be milked and fattened for the kill.

Humans learn not only intuitively and by example. They learn by communicating their ideas. But humans have inborn filters that prevent perfect communication.

And once again Thomas Aquinas was ahead of the game.

Aquinas intuitively understood that the transmission of information and insights involves yet another triple reality, involving the sender of the message, the receiver of the message, and the message itself. And he figured out the corollary to that. This insight or discovery, which explains the dynamic between sender and recipient of a message, is described in the next section.

It constitutes a startling premonition of modern psychology.

Thomistic Philosophy Meets Developmental Psychology

Aquinas's second great insight added an important nuance to empirical thinking on human communications, which assumed that humans were capable of absorbing and processing reality in a totally objective way. He based his insight on the Hebrew tradition, which begins with a story of *homo sapiens* losing his objectivity in making an important decision (or perhaps we can say an unimportant decision with grave consequences).

The decision to eat an apple from a tree forbidden by G-d began a string of irrational decisions by humans – as viewed from the Jewish and Christian value system. After the alleged revelations in Genesis, Jewish and Christian philosophers saw in themselves and their fellow humans a tendency to distort reality to suit the purposes of the observer. They referred to that subjective distortion as "sin" and concluded that it was an inherited trait, passed on by our first ancestors.

By the time that Aquinas came into the scene, the idea that humans were not objective in their processing of information was well ingrained. With that extraordinary prescience that characterized him, Aquinas was able to proclaim an elementary principle of communications, saying: "Whatever is received is received according to the mode of the receiver."

As Richard Rohr explains, we now call this a principle of developmental psychology. He elaborates on the idea, which he calls "one of the most helpful and clarifying elements for the modern mind," i.e., that it understands things in terms of *developmental stages*. Continues Rohr:

> *As a preacher and teacher, I know that I can say one thing and it will be heard on as many as ten different levels, depending upon the inner psychological and spiritual maturity of the listener.*

Rohr follows that up with a quaint anecdote about how the lowly and the mighty react to the same homily. He says:

> *I can give what I think is a lousy sermon, yet a humble woman will come to me after mass in tears of gratitude for the beauty of something that spoke to her deeply. She may not be highly educated, but she is spiritually evolved. Another 'smart' but cognitively rigid person will hear the same sermon and is only convinced that I am a heretic. Mature people can make lemonade out of lemons. Immature people can turn the sweetest lemonade tart and sour.*

The trick, of course, is to mature into adulthood and then continue maturing so as to be able to appreciate the goodness in others, inspire them with our own goodness, and grow together as a family, tribe, community or the greater society. To do so, humans need to re-program themselves to eliminate the excessive subjectivity that so often leads to doing harm to themselves and others.

The question is how to do this, as well as whether some sort of religious "redemption" is needed.

Science and the Idea of Redemption

In the chapter that follows, we will seek to determine the historical and rational roots of what Christians call redemption and we can refer to as "reconnection" or "reprogramming." We will do so in an effort to determine whether the whole idea is compatible with the latest pronouncements of science.

Of course, there are two kinds of redemption, according to the model of Christianity we are considering. One, which goes back all the way to the apostle Paul, was radical, to put it mildly. It clearly fits in with the Valley of Tears model. Under it, Christians were required to "die" to themselves and be reborn "clothed in Christ."

Clearly, St. Paul took literally the metaphor of the mustard seed, which must "die" when planted in order to sprout into a tree. I think St. Paul read too much into the parable in question. Jesus was saying that, in order to grow, we must look less to our own needs and more to those around us.

Once again, C.S. Lewis said it best: "True humility is not thinking less of yourself; it is thinking of yourself less." In other words, the mature human who wants to achieve full human fulfillment, does not have to wallow in the mud, but does have to be willing to lift fellow beings out of the mud. He has to be willing to "pay it forward," as per the modern movie of that title.

Life is all about joy, but a good measure of selflessness helps enormously to achieve joy. *Redemption does not require dying to yourself, but it does require taking notice of those around you who are dying of thirst, or hunger or loneliness.*

Before leaving this topic, I offer my reader a passage from one of the most radical followers of the valley of tears model, who (interestingly and probably not coincidentally) incorporates Aquinas's triune description of the human mind.

His name was Ignatius of Loyola.

Ignatius of Loyola Describes Total Immersion in God

Ignatius of Loyola is one of the figures of Catholic history that is most revered. He founded an order of priests (the Jesuits) that counts the largest number, by far, of Middle Age and Renaissance scientists. Theologically, Jesuits were the Catholic answer to the Protestant Reformation.

They cherished intellectual accomplishment, as well as spiritual purity. Ignatius was anything but wine and roses; he might have frowned on his most recent and best-known follower, Pope Francis, and his tiptoe-through-the-tulips style. Ignatius's best-known invention, other than the Jesuit order itself, is what are called "spiritual exercises," which are all-absorbing meditations where the faithful spend at minimum three days contemplating the mysteries of creation, death and a presumed eternal life in the company of the almighty.

Like other religious mystics, Ignatius conceived of more than a reconnection between man and God; he conceptualized a total merger, a confluence of all human traits with God's. If being a practicing Christian were like being an athlete, his vision would be the "extreme sport" of Christianity:

> *Take, Lord, and receive all my liberty, my memory, understanding, and my entire will. All I have and call my own. You have given all to me.*
>
> *To you, Lord, I return it. Everything is yours; do with it what you will. Give me only your love and your grace; this is enough for me.*

Even this extreme merger of the human will with the divine will does not result in some sort of self-imposed death, as preached by St. Paul. It can be interpreted to mean no more and no less than a scientifically sound effort to control our urges and to love our neighbors as much as ourselves. As such, it is a very sound approach to both personal and societal challenges.

And that is in line with the theme of this book, which argues that the bible, as a whole, provides solid life lessons. As natural theologians – as scientists – we can readily, rationally accept them. We can reason our way to the conclusion that some form of reconnection, some rewiring of the human hardware, needs to take place. But why does the exercise have to involve a God-man, or death on a cross?

We will seek answers to that question, as well as to a more elementary one, which arises from a totally different hypothesis; it is one that would have been unthinkable to the Greeks, as well as to scholastics like Aquinas.

Premised on that idea that *homo sapiens* is not essentially different from the animals, many anthropologists and evolutionary biologists argue that humankind has no rational power of decision. A great deal of academic discussion nowadays is thus devoted to the rather elementary question of whether free will is a figment of our imagination.

The next chapter tackles that question, as well as the consequences that flow from concluding, with the ancients as well as with all mainstream religions, that humans not only have free will, but the moral responsibility to exercise it towards the betterment of themselves and the earth they inhabit.

Chapter III

FREE WILL AND THE MODERN SOCIAL SCIENCES

> *'If you go on denying that I am free in my decisions,*
> *I'll punch your nose.'*
> *The old man gets red in his face. 'I deplore your*
> *unpardonable behavior.'*
> *'I apologize. I lost my temper.'*
> *'You really ought to control yourself.'*
> *'Thank you. The experiment was conclusive."*
> Arthur Koestler, *The Ghost in the Machine.*

The above fictional dialogue, between student and teacher, is as compelling on the existence of free will as any that I have ever read.

Koestler's book, quoted above, is a more methodical (and obviously more complete) treatise tending to prove that humans do have the power of decision – that we don't simply respond to stimuli like rats in a laboratory. He was doing his best, in 1967, to debunk the newfangled psychology theory known as "behaviorism."

Koestler referred to that model of human behavior as being "ratomorphic," because advocates of that theory seek to convince us that humans are really just more sophisticated rodents. Perhaps our powers of response to stimuli are more convoluted – i.e., they have many more variables than those of rats. But they are not all that different.

Like animals, the behaviorist argues, we humans are controlled by our appetites as we try to win the proverbial rat race. Perhaps they take seriously the popular saying, "whether we win or lose the rat race, we all remain rats."

Ratomorphic/behaviorist theory reached a peak of popularity with the writings of B.F. Skinner, who did his best to convince us that humans really did act like rats. But the theory has fallen from grace – particularly among clinical psychologists and psychiatrists, who have an actual stake in curing people, as opposed to writing academic papers.

Most social scientists and almost all lay people simply assume that humans have some measure of free will, as well as an inner voice that guides our species in a noble direction. They may not refer to it as such; they may not use terms like a soul to describe the part of our being that struggles with our simple appetites to convince us to be virtuous, rather than gluttons.

Let's review the history and see how philosophers, theologians, historians, and authors of both poetry and prose have pondered, described and otherwise explicitly recognized the existence of an inner-voice or sub-conscious inside the human mind-psyche.

Emerson's "Inward Voice"

A whole glossary of terms has been used for the supposed light that guides our decision-making and which some have called a soul, others a conscience, still others a "moral barometer" or the "moral high-ground."

The American poet Ralph Waldo Emerson called it an "inward voice" that we should follow, rather than any other voice. Because, he said, it means following "yourself."

But what is "yourself," if not the sum of your appetites, of your instincts? Is it your reasoning mind? Is it a combination of conscious and sub-conscious, of dreams and ambitions that lead you higher and higher, or of frustrations and fears that lead you to lower, perhaps destructive and egotistical behavior?

In modern political discourse, we borrow a maritime term and refer to a "moral compass." In sociology and criminology, we often refer to the importance of norms that prohibit "anti-social" behavior. Our entire system of laws is premised on the idea that humans can adjust their behavior by volitional acts.

Humankind's ability to modify conduct is best exemplified in the criminal law.

Behavior Modification in Criminal Justice

Punishing people for violating the law is said to have at least two principal rationales: one is "retribution" and the other is "rehabilitation." The former is controversial, as a significant number of thinkers and policy-makers argue that "retribution" is a fancy name for "vengeance" and has no place in our legal code.

But everyone agrees on the value of rehabilitation of a convicted criminal, particularly where small offenses are concerned. We put criminals on parole, probation and every other kind of conditional liberty in the hope that they will exercise that liberty better in the future than in the past.

The entire science of economics is premised on people choosing to follow their "enlightened self-interest" in their consumption and production choices. Saving for a future "rainy day" is an important component of capital formation. Choosing to band together into labor unions and bargaining collectively rather than individually is a way to optimize future scenarios.

In recent years, the entire new field of Decision Analysis relies on sophisticated quantitative models to select from options such as whether to dig or not dig an exploratory well, engage in expensive research and development of a new drug, or choose from among various transportation technologies.

While practitioners of psychology, economics and criminal justice go about their business assuming some measure of free will, theorists struggle with whether free will is real, and if so, where it comes from.

The problem is that there is no obvious organism in the human body where one can pinpoint the power to make decisions. Our senses clearly pick up stimuli of all sorts, which are transmitted to the brain by the nervous system; from there the brain reacts and sends signals to other parts of the human body, such as muscles, glands, and vocal chords, causing movement and communication.

Modern biology can explain all of that fairly precisely. But there is a "ghost inside the machine" that causes some of us to react differently than others, when subjected to the same stimuli. Even identical twins develop different personalities, despite having the same genetic make-up and same environmental experiences. (The book to read is *No Two Alike*, by Judith Rich Harris.)

So where is the organism that emotes – that chooses to share or not, to cheat or not, to be cruel or kind? The question baffles scientists as much today as it ever has.

The mystery is so great that for the last six-thousand years, or the entire time since humans have been able to put ideas into writing, the organ identified in prose and poetry as the one that emotes is the heart. From that literary notion we get expressions such as "heartless" and "disheartening" and famous names such as Braveheart and Richard the Lionheart.

Many modern scientists dance around the question and ultimately move on to their particular occupation, making some meek pronouncement that suggests a vague acceptance of the notion of free will, but no clue as to the source or nature of it.

Here's a good example, from Italian theoretical physicist, Carlo Rovelli, author of the international bestseller, *Seven Brief Lessons of Physics*:

> *There is one issue in particular regarding ourselves that leaves us perplexed, what does it mean, our being free to make decisions, if our behavior does nothing but follow the predetermined laws of nature? The solution to the confusion lies elsewhere. When we say that we are free, this means that how we behave is determined by what happens within us, within the brain, and not by external factors.*

Rovelli may be an acclaimed theoretical physicist; but he has not really cleared the confusion, other than to say that the human power of decision comes from within the body and not from outside.

Well, we knew that!

It is beyond my capacity, as well as the scope of this book, to determine where exactly the human power of decision finds its source. It is fair to say that there is no consensus among academicians about the inner workings of human free will. A fairly straightforward definition comes from German-American political scientist, Alexander Wendt of Ohio State University, who offers this definition:

> *Will is the essence of agency, a power to animate and move the body – and the mind in the form of attention – from the essentially passive stance of Cognition to active, purposeful engagement with the world.*

Note that Wendt contradicts Harari when he refers to "purposeful" conduct, in contradistinction to Harari's previously quoted insistence that human actions have no purpose.

In the end, my reader must make a choice: either to assume that we make choices or to assume that we don't – that we are automatons who simply follow our instincts. If you assume the latter theory, you should probably put this book down and engage in some other, perhaps more trivial pursuit.

Otherwise, let's continue with a quick review of what prominent psychologists and sociologists say about those little, seemingly autonomous, invisible yet ever-present voices that seem to be vying for attention inside our consciousness.

The unquestioned pioneer in this area of science was Sigmund Freud.

Freud's "Super-Ego"

Freud refers to the "super-ego" that should guide the "ego" and the "id," which make up the "self." In Freudian terms, the self is what Dr. Scott Peck calls a "conglomerate of psychic components" that define a person's genetically determined temperament" and "accumulated experiential learning." These are roughly what psychologists nowadays commonly refer to as "nature" and "nurture," two components which in combination are thought to define the human personality.

As we shall see, the combination of nature and nurture, of our genes and our social experiences, does not fully define our species. There is something deeper, something more mystical, something that draws on an inbred need to transcend our appetites.

This intuitive reality has traditionally been referred to by philosophers and theologians as the "soul" and by writers and poets as the "heart." Modern psychology has fine-tuned the idea of a soul; and even Freud's close disciples, beginning with Carl Jung, seemed to recognize elements of the human psyche that respond to stimulants from outside the observable components of the body.

Jung's "Shadow" of Evil

Jung, for example, modified Freud's concepts and argued for the existence of an "unconscious" element in the self, which (as John Freeman explains) "is the great guide, friend and adviser of the conscious." Jung concocted a novel term for what most of us refer to as "bad" or "evil." He called it the "The Shadow," or that compartment in our mind which contains things we would rather not acknowledge. One's "refusal to meet the Shadow" is in essence a pathology that is at the root of human evil.

Here's how the previously cited popular psychiatrist, M. Scott Peck, describes the almost universal phenomenon:

> *Most of us, when pushed up against the wall by evidence of our own sins, failures or imperfections, will acknowledge our Shadow…By his use of the word 'refusal,' Jung was implying something far more active….Those who are evil refuse to bear the pain of guilt or to allow the Shadow into consciousness and 'meet' it.*

Ultimately, the bad or evil persons among us (as Peck caricatures them) "set about – often at great effort – militantly trying to destroy the evidence of their sin." They even fall as low as to destroy not only the evidence of this misconduct but even "anyone who speaks of it or represents it." And, continues Peck, "in this act of destruction, their evil is committed."

Jung's disciple, Marie L. von Franz provided us with a historical survey of the human realization that there is good and evil, as well as an internal moral barometer that should help humans distinguish between one and the other.

Von Franz's "Inner Center"

Jung's most celebrated disciple was Marie L. von Franz. In her writings, she refers to an "inner center" and argues that "throughout the ages men have been intuitively aware of the existence of such an inner center." And she continues:

> *The Greeks called it man's inner 'daimon'; in Egypt it was expressed by the concept of the 'Ba-soul'; and the Romans worshipped it as the 'genius' native to each individual.*

Von Franz based a lot of her analysis on a native tribe called the Naskapi Indians, who still exist as simple hunters in the Labrador Peninsula and who "live in isolated family groups, so far from one another that they have not been able to evolve tribal customs or collective religious beliefs and ceremonies." She describes in vivid terms the values espoused by the Naskapi:

> *In his basic view of life, the soul of man is simply an 'inner companion' whom he calls 'my friend' or "Mista'peo," meaning 'Great Man.' Mista'peo dwells in the heart and is immortal; in the moment of death, or shortly before, he leaves the individual, and later reincarnates himself in another being.*

As in the bible, where Jesus admonished his followers to stop wanting to remove the speck in your neighbor's eye, while not noticing the beam in your own, Von Franz describes each person as denying in himself "those qualities and impulses" which she "can plainly see in other people."

And she proceeds to offer us a rather familiar list, which she calls "little sins" and which includes "such things as egotism, mental laziness and sloppiness; unreal fantasies, schemes and plots; carelessness and cowardice; inordinate love of money and possessions...."

The point is we have a natural sense of morality. We know sin when we see it.

Original Sin and Modern Science

As a good engineer, I felt I had to try an empirical analysis of the human flaw by recurring to a concept in the field of thermodynamics. I had read about "sociological entropy," or the idea that, absent a force from outside the system, entropy (disorder) increases with time. Perhaps, I mused, the reality of entropy can be applied not only to inorganic elements, but also to human societies. It was precisely in that vein that the term sociological entropy was used.

I applied the same idea to individuals, as we relate to God, and came up with "theological entropy," to describe the disorder that ostensibly followed the disconnect described in the Adam and Eve story.

The more I think about it, the more logical it sounds; it also accords with what we know of our ancestors and what we think we sense of a moral code that is implanted inside our beings.

Let's follow the logic.

We have already concluded that our species was most likely not created to have a deep flaw that constantly forces us, if we want to get along with our spouses, friends and work colleagues, to say we are sorry. (I am reminded of the illusory and somewhat contradictory saying in a popular movie of the 60's which suggested that "Love is never having to say you're sorry.")

Logically, if we were created by a perfect God, you would think we would not have to be constantly forgiven by our life companions for following our instincts. Animals don't have to apologize for following their instincts; their nature is not ambivalent; they have no character flaw.

Humans definitely do. And that is why parents notice that their kids need to be disciplined, and teachers, social workers and parole officers acknowledge that human beings often act in an anti-social way. But is it an inherited trait, as portrayed in the biblical account?

As it happens, the modern science of psychology is quite compatible with the otherwise mythical story of Adam and Eve. One of the most popular restatements of that story, using clinical terms, is contained in the writings of M. Scott Peck, M.D.

Scott Peck: The Loss of Innocence and the Acquisition of Consciousness

In his 1997 best-seller, Dr. Peck draws a very compelling parallel between biblical notions of sin and modern psychiatric insights into what appears as a constant, internal conflict between certain human instincts and the nobler, more rational and mature norms of conduct prescribed by law and custom.

Here's a sample of what Peck considers the confluence of Genesis with psychology:

> [T]*he first result of the Tree of the Knowledge of Good and Evil is that Adam and Eve become shy or modest because they are now 'self-conscious.' They are aware that they are naked. From this we can also extrapolate that the emotions of guilt and shame are manifestations of consciousness, and although both emotions can be exaggerated to the point of pathology, within limits they are an essential part of our humanity and necessary for our psychological development and functioning.*

Peck then proceeds to describe the difference between genetic traits that our species possesses and developed (evolved) traits that have been acquired in the effort to impose intellectual discipline on instincts that often seem out of control. This process, which Peck characterizes as leading to "maturity," happens as humans "arrive at the realization that consequences follow actions." In other words, when we stop "acting at a purely reflexive or instinctual level" and become "conscious that we have free will."

The reader will note that the idea of free will is a recurring theme in this book. Because it is not accepted in many parts of academia, it is necessary to our logical construct of a scientific God to consider alternative theories of human conduct that don't include free will. One of those theories, which goes by the name of "sociobiology," posits that human actions are the result of biological reflexes over which we have no real control.

Sociobiology

As mentioned before, modern psychologists analyze human behavior in terms of two principal factors: "nature" (genetic composition) and "nurture" (environmental factors, including parental upbringing and peer pressure). For a couple of centuries now, psychologists have studied and applied these two factors, but have never foreclosed the possibility of a third factor that somehow interplays with those two and that involves free will.

The possibility of a third element in human conduct was suddenly foreclosed by a theory that is only a couple of decades old.

About a quarter of a century ago, a school of evolutionary biologists formulated a theory that merged Darwinian natural selection with human behavior analysis and thus foreclosed any possibility of volitional action. It was pioneered by Edward O. Wilson, at Harvard, who referred to his theory as "sociobiology." Because it absolved the individual person from moral responsibility for antisocial behavior, when Wilson first proposed his ideas at a professional convention, they threw a bucket of cold water on him.

I will do the same here – though figuratively. To me, it is self-evident that humans have free will, and that most of us can control our instinctive impulses enough to do the right thing most of the time. Notwithstanding what Wilson and his cadre of "sociobiologists" might argue, humans are not automatons or machines.

We will never be replaced by computers.

Recent studies in sociology point in the opposite direction from

Wilson's mechanistic view of man. Not only is it apparent that our mind-psyche is a separate, though interacting dimension from the purely physical brain, but it is quite probable that a third dimension is at play.

That third dimension is less tangible and more ethereal, but no less real. It makes its presence known in studies of individuals who are religious believers. How the mind-psyche interacts with this spiritual dimension in man – how we seem to thrive when we act according to a moral barometer – is the subject of the next few sections.

The interconnectedness of perceived moral principles and human psychology is particularly apparent when we analyze certain kinds of anti-social behaviors. The reader will note that science uses terminology that is totally different from classical Judeo-Christian terms. Yet the concepts are quite similar in their application.

Let's delve into that interconnection; as we do, I expect that my readers will be surprised to that many behaviors which religions have long condemned as immoral have an equivalent in psychology.

Sin hasn't left us; it has acquired a new name.

Sin in Contemporary Social Science

What most religions refer to as "sinful" behavior, social scientists (psychologists, sociologists, criminologists) refer to it as anti-social behavior. It really means the same thing, though it is applied to a smaller range of sins.

Let's use the Ten Commandments of the Hebrew bible as a handy guide to distinguish what science and theology might consider improper conduct. (I leave out of consideration the strictly religious prescriptions such as those that prohibit invoking the name of God or require worship on the Sabbath.)

Killing innocents, stealing from others and giving false witness against one's neighbor would be forbidden by both the religious and the lay codes. Coveting or engaging in sexual contact with another person's spouse is no longer forbidden in countries that are not Islamic; but it is still seen as a negative trait by courts tasked with having to determine primary custody of minors.

Honoring one's parents is not a societal norm, *per se,* but society does give a lot of authority to parents to enforce rules of conduct inside the home, including not inflicting pain on siblings or destroying property. The civic society will also enforce parental rule-making that requires minors to avoid "adult" beverages and movies and to engage in some form of schooling.

In summary, it is evident that social scientists (and societal lawgivers) enforce their concept of the natural law; they also promulgate human laws needed for those not yet considered mature enough to drink or gamble or view pornography. Transgressions of those laws are not called "sins," but the effect is quite similar. In one case the transgression removes the violator from the lay community and in the other the transgression removes the violator from the community of those with entry to the Promised Land.

Repeat violators, as we shall see, merit even harsher treatment. Scientists call it "addiction."

"Addicted" to Sin

Modern scientists shy away from any description of anti-social behavior that puts the blame on sinfulness, as prescribed in the classic Judeo-Christian scheme. But those who practice the healing crafts, in the real world, still find room for the concept of God. In effect, scientists shy away from bringing up God when diagnosing a *problem*; and yet allow God to seep into the conversation when crafting a *solution*.

Belief in a transcendent being has turned out to be almost a necessity in the modern world – particularly when dealing with addiction. Philosopher Richard Rohr is one of many who perceives an undeniable correlation between addiction (understood to mean the inability to control self-destructive behavior) and the Christian concept of an infinite, benevolent force that acts to liberate the addict who is otherwise enslaved by his or her instincts.

In this context, there is no denying the positive impact of Bill Wilson's "Twelve Steps" method of treatment for alcohol and drug addicts. There are skeptics, of course. But the proof is in number of addicts who have regained a level of functionality; almost all admit that -

> *On the practical (read "transformational") level, the Gospel message of Jesus and the Twelve-Step message of Bill Wilson are largely the same. The first foundational connection is that addiction can be a metaphor for what the biblical tradition called sin. It is quite helpful to see sin, like addiction, as a very destructive disease, instead of merely something that is culpable, punishable, and "makes God unhappy." If sin indeed makes God unhappy, it is because God loves us, desires nothing more than our happiness, and wills the healing of our disease.*

More and more, the social sciences are accepting the idea that there is a right and a wrong way for humans to conduct themselves.

In the law, the test of whether a mentally ill person is able to stand trial is based on his ability to tell right from wrong. In psychology, we refer to those who fail to respect the rights of others as being "dysfunctional."

In politics, the terminology reverts to more conventional terms. We say that the politician who acts less selfishly (or less beholden to special interests) has the "moral high ground."

Even in economics, which is the most quantitative of the social sciences, the terminology of ethics creeps in. For example, economists argue that the system should not reward the person who violates the rules – that kind of reward for unvirtuous behavior is referred to as a "moral hazard." The best example of a moral hazard is to bail out (reward) a company that has speculated with the assets of its investors and come close to bankruptcy due to its recklessness.

All of those are examples of the acceptance of a human volitional capacity, as analyzed and applied by the various social sciences. But it is in psychology that we find the definition of an illness that comes closest to the theological concept of sin, based on repetitive, remorseless, antisocial behavior.

Modern science calls this person a sociopath.

The Sociopath Next Door

In my view, some people diagnosed as sociopaths have developed that condition without having any genetic or environmental component; their dysfunction is simply a result of responding to their passions so repeatedly that they lose all ability to align their free will with the dictates of morality.

Psychologists describe people like that as having no remorse from constant – and often damaging – lying. (The book to read is *The Sociopath Next Door* by Martha Stout.)

It is said that the condition affects as many as one in 25 Americans.

What psychologists call sociopathic behavior is easy for us to recognize. Anyone who is a teacher or counselor has encountered the classic case of a parents who lie about their child's supposed dysfunction. They'll say, for example, that their child was removed from one school because he/she was underperforming. When a little more digging is done, it turns out that the exact opposite is true: The child was performing well before being removed to a school that the parents chose – typically for their own convenience.

Constant lying for selfish reasons leads to a condition of losing all sense of guilt. You convince yourself that you're lying for a greater cause, as might be done by a prisoner of war or a hostage negotiator. Except that in this case it's done for less admirable reasons.

The condition, referred by most of us as "selfishness" is defined by psychologists as "narcissism," which they consider pathological. Here's how Harvard Psychiatrist Martha Stout defines it:

> *Narcissism is, in a metaphorical sense, one half of what sociopathy is. Even clinical narcissists are able to feel most emotions as strongly as anyone else does, from guilt to sadness to desperate love and passion. The half that is missing is the crucial ability to understand what other people are feeling. Narcissism is a failure not of conscience but of empathy, which is the capacity*

Science and the Theory of God

to perceive emotions in others and so react to them appropriately.

The clinical narcissist become a sociopath when she lies and exaggerates in order to justify her narcissism. The pathological stage is achieved when there is no remorse for covering up the failure to fulfill parenting responsibilities or other societal obligations.

M Scott Peck, M.D. called such individuals "people of the lie" whose condition is beyond sickness. He explains the difference:

> *Many of those who are 'sick' deal with their venom internally, turning it painfully upon themselves if they choose not to seek help. Those who are evil go another way. They fail to suffer. Because they lash out at others and use them as scapegoats, it is the people around them who must suffer.*

Lying to cover one's transgressions, and even worse blaming others, is what most of use call immoral behavior. We sense that it is culpable conduct that must be redressed somehow, whether in this life or in a hereafter. We conclude, following instinct (and logic) that if there is a God and an after-life, there has to be some behavior modification, accompanied perhaps by some restitution to those you have harmed.

The Spiritual Black Box and Its Empirical Analogy, the Black Hole

Whether we rank or do not rank the sins as "little" or "big," there seems to be a consensus on values: a rather undeniable and complete catalog of ethical norms to which the civilized world generally adheres. And here I am not referring to the religious world or even to those who believe in an almighty being.

Humans of all stripes and beliefs, of all eras and cultures, have recognized an inner code, an inner voice that distinguishes right from wrong. The secular society – particularly academia – struggles with the concept. The great majority of university professors, in almost all disciplines except philosophy and theology, start from the *a priori* assumption that human beings are simply animals with a more sophisticated brain.

Yet there is a trend away from that simple assumption, and it is not just among psychologists. Another science that is trending towards a recognition that there is a spiritual dimension at play is sociology.

The principal methodology of sociology is statistical analysis. And every single study that I have read about indicates a high probability of something intangible, something mystical, something that cannot be measured directly, but whose indirect consequences are undeniable. The typical study concludes that there is a high correlation between faith and psychic health.

The problem is that no one is able to penetrate the human psyche and determine what element in that impenetrable "black box" is causing the improvement in psychic health. Empirical science posits models of a spiritual dimension, but lacks the tools to measure and confirm its existence.

It reminds me of the phenomenon that can be detected at the boundary (or "event horizon") of a black hole. These are very dense concentrations of matter which appear at the middle of many, if not most, galaxies, and which are so gravitationally powerful that light cannot escape from inside.

So we cannot see them. We cannot see black holes, but we can sense they exist, from observations that happen just on the outside. One fairly easy phenomenon is the effect the black holes have on stars in their vicinity. We see them spinning rather rapidly as if they were droplets of water going into a drain that sucks them in.

Thus it happens with sociology, when its practitioners analyze the spiritual realm. Even the most secular academicians sense that there is an intangible reality at play. They sense that it is important and that it is right there, as a big magnet, beckoning mankind to behave. They can't measure the actual phenomenon; but they can measure its "event horizon." Somewhere beyond the measurable, palpable, solid reality of an enormous physical being that propelled galaxies spinning and obeying all kinds of physical laws, there is a more primeval reality

Why do I say that? Why am so sure? Well, for one thing, because it is the most plausible explanation of phenomena that can be measured in statistical correlations, even if the empirical cause-and-effect is not perceived.

In the end, we are required to make conclusions based on that is the *most plausible explanation, applying the standard of juries in civil cases.*

Fair-minded scientists are able to discern such truths even at the Mount Olympus of American academia, called Yale.

God and Woman at Yale

A recent symposium at Yale, involving no less than 30 faculty members and directed by my niece, Margarita Mooney, gives us a sample of the kind of dialogue in which academicians engage, while struggling to define this mythical, intangible aura that envelopes the human brain and gives it a resonance that is somewhere beyond the strictly material realm. This particular group of Yale professors recognize that there is some sort of internal conversation going on, as if the brain were consulting the "heart" or "soul" or "conscience" on the meaning of human existence; my niece quotes Oxford:

> *How far does cognitive science take us in understanding existential questions? 'The human brain is not exclusively a thought-making machine, but rather an innate channel for dialogue with a conscious, surrounding reality.' (Oxford Handbook of Spirituality and Psychology, p. 2)*

The engineer in me resists, as too vague and amorphous, the idea that the existential conversation is between the brain and a "conscious, surrounding reality." But I do like the image of an "innate channel" for dialogue with some other, perhaps non-material "reality." Where exactly that non-material reality lies in our bodies is a question for another day; what is clear is that we humans share at least this one trait with God; and that we are quite different at least in that sense – from higher-order animals like a chimpanzee.

Our hardware resembles that of the animals. But our software resembles God's.

As a side note, I should point out that Margarita Mooney and I have resources which were not available to the A-team. We have had the benefit of Yale and Harvard scholarship, contained in amazing deposits of learning. She has Sterling Library, with four million volumes. I had Langdell Library (with 18 million volumes, though spread among many different buildings).

We both now have the Internet, which is an almost inexhaustible library.

The above begs the question: How was it possible for Aquinas, bereft of those extensive sources, to combine in one person the kind of enlightenment that made it possible for him to grasp concepts as advanced as the analogy of man to computer?

And a related question: Can our species collect enough wisdom, absorb enough attitude-adjustment therapy and acquire the required level of maturity, individually and collectively, to ultimately conquer disease, institute societal systems that allow for dispute resolution without resort to violence, and eventually achieve universal welfare for all humankind?

Or is it likely, as some prophets of doom have opined, that humankind has reached a critical mass of collective mischief that will lead to unavoidable mass extinction? Is it too late for redemption?

The following chapters in this book examine that question, beginning with the very validity of the concept of redemption, as viewed by science.

Chapter IV

FALLEN MAN TO RENASSAINCE MAN

> *Your ancestors' lousy childhoods or excellent adventures might change your personality, bequeathing anxiety or resilience by altering the epigenetic expressions of genes in the brain.* Dan Hurley, *Discover Magazine.*
>
> *I believe we are all worms. But I am convinced that I am a glow worm.* Winston Churchill

Science, as discussed in the prior chapters, generally supports the idea that our species needs some form of rehabilitation. The original owner's manual does not seem applicable any more. The contraption that is our body and mind needs major fine-tuning.

Social scientists call it "behavior modification" and it starts from birth, as if the bad habits were inherited. The child has to be taught to be sociable, to share with other children. The adolescent has to be guided towards a proper understanding of sexuality; it is called "sex education."

Adults have to be encouraged to follow all kinds of laws and customs, even when the authorities are not watching. Constraints on the most outrageous human behavior are enforced by the civil law, which can take the form of "tort" suits that make the "tortfeasor" pay for his/her negligence. Or they can take the form of criminal penalties, which separate the transgressor from society or even impose the death penalty.

All of us humans, of every age, need both internal and external restraints that channel our actions towards the measured, disciplined borders of the common good, rather than the excesses that tempt us to overflow into riverbanks of anti-social behavior.

In other words, it can be stated with a fair degree of scientific certainty: Humans are wired to function in one way, yet suffer collectively from a virus that pollutes, contaminates, derails and otherwise propels us in a disordered way. *We are the only species that has to unlearn something that got into our system, at some point and in some way.*

Modern science struggles to give clinical names to deviant human conduct. In the prior chapter, we discussed the case of sociopaths, whose personality disorder is akin to what in the past would have been called an inveterate liar. A little bit less blatant is the failure of judgment that we characterize as bias or prejudice, which clouds our judgment and impedes objective reasoning.

And everyone knows that humans tend to be self-centered and vain to the point of absurdity; how often do we notice a speck in our neighbor's eye while ignoring a much bigger beam in our own?

Of Pride and Prejudice

The Genesis story concludes with an act of disobedience; Adam and Eve are enticed to climb the existential ladder and become like God in one fell swoop. For that they earn a lifetime of hard labor – from history's all-time life of ease to what the biblical writers call a valley of tears.

The scientific name for this flaw in our personalities is "narcissism," which is a term adopted from the tale of Narcissus, who was so impressed with himself that he spent considerable time admiring his own reflection.

It is hard to argue with the idea that excessive pride is the cause of much human misery. How many wars were caused by the pride of a king, an emperor or a sheik?

Ethnic and national pride can get so out-of-control that it leads to riots during sporting events.

Parental pride can be just as deadly. It is a common experience to have fistfights between parents of opposing Little League baseball teams. Referees and umpires in all sporting events are the objects of verbal attacks, as well as an occasional projectile hurled by a fan of the team that feels aggrieved by a close call.

All humans struggle to moderate the urge to control the conversation, to outshine the competition even when there is no competition. The ability to listen to others of our species is such a prized talent that it enables those with a psychology degree and a couch to charge substantial fees for the practice. For the rest of us, who are not professional listeners, it is an important acquired trait.

The converse of pride is prejudice, and it is equally devastating. Humans have a tendency to engage in intellectual short-cuts, reaching conclusions that are not supported by the data. The flaw caused by this tendency has a scientific name.

The study of bias in our analytical thinking goes by the title of "heuristics."

Heuristics is a field which studies how humans are predisposed to some prejudice or bias that clouds our judgment. Typically, our judgment is clouded towards our own self-interest, our own short-term welfare, rather than the collective or long-term view of things.

Humans are instinctively narrow-minded; we are wired to "miss the forest for the trees." The flaw in our nature is particularly evident during the developmental stage of the *homo sapiens* child.

Developmental Disorders: The Role of Nature and Nurture

Certain human disorders seem curable, either by psychoanalytical therapy or by psychiatric drugs. In their effort to remedy developmental defects in the growing child, scientists struggle to identify their cause. Is it the faulty wiring of the hardware (nature) or the faulty programming of the software (nurture)?

For example, a child who perceives himself to be deficient in a physical trait (e.g., height, or intelligence) can fail to bond with the parents. Consequently, he suffers from low self-esteem. The condition is nowadays referred to by psychologists as "separation anxiety depression."

Is the low self-esteem caused by the perceived physical deficiency (nature)? Is it caused by the failure of the parents to give the child the love and acceptance he craves for (nurture)? Or is it a combination of both?

The nature-versus-nurture causal distinction is easier to diagnose when the faulty hardware is the result of willful abuse or neglect by the parents. Sexual abuse by a stepfather, teacher or nanny can lead to all kinds of deviant behavior in the expression of sexual appetites. Physical or mental abuse from parents or proxies can lead to criminal behavior. In those cases, it seems like the volitional act of one generation causes permanent, inherited damage to the next.

Deviant *nurturing* causes permanent damage to *nature*.

The scientific debate over character flaws and whether they are totally caused by "nature" as opposed to "nurture" is by no means subsiding. Adding to the complexity is a whole new branch of genetics, buttressed by the discovery that personality traits (good and bad) can be acquired in one generation and passed on to the next.

It is called "epigenetics."

Epigenetics and Inherited Traits

Until very recently, science has rejected any possibility that traits acquired from parents during their lifetimes could be passed down or inherited by their offspring. Darwinism pretty much displaced that notion, which was referred to as "Lamarckism" for its founding theorist, Jean Baptiste Lamarck. The idea that our ancestors could have passed on to us DNA that they themselves were not born with was anathema to scientists – considered as infantile as the idea that the sun revolves around the earth.

But that idea has recently begun to flourish with the discovery of "epigenetics," which is a modern variation of classical genetics that seems amply supported by the evidence. Simply put, it means that the sins of the parents (drug abuse, bad dietary habits, domestic violence) can be visited on the children – and maybe on their children.

Dan Hurley explains it in a recent edition (originally appearing in May, 2013) of *Discover* Magazine, devoted entirely to genes and health:

> *Like silt deposited on the cogs of a finely tuned machine after the seawater of a tsunami recedes, our experiences, and those of our forebears, are never gone, even if they have been forgotten. They become a part of us, a molecular residue holding fast to our genetic scaffolding. The DNA remains the same, but psychological and behavioral tendencies are inherited. You might have inherited not just your grandmother's knobby knees, but also her predisposition toward depression caused by the neglect she suffered as a newborn.*

If psychological and behavioral tendencies are inherited, as modern experiments suggest, then it is not farfetched at all to believe that a decision by our earliest ancestors to contaminate themselves or their environment could have deformed our otherwise pristine genetic composition. If the experiences of our forebears "are never gone, even if they have been forgotten," then it is quite conceivable that their negative experiences have cast a stigma that lives with us in some way.

Epigenetics is not settled science, by any means. It has not been tested past the second generation in humans. In fact, as I was putting the finishing touches on this book, there appeared an article in Harvard Magazine (May-June 2017) that quoted Dr. Karin Michels, head of UCLA's Fielding School of Public Health, for the proposition that "no researcher has proven transgenerational epigenetic inheritance."

If it becomes settled science, it will tend to support the biblical account that we have inherited a tendency to follow our disordered instincts, those out-of-control appetites that, in limited measure, are part of our nature, but in our current "sinful" (scientifically flawed) state, cloud our will to the point that we go counter to the will of our creator (scientifically known as the Big Banger).

Using the vernacular, we would say that we have a tendency to "go against the better angels of our nature."

Philosophers would say we have a slight but discernible inclination to choose an immediate or short-term good even though we know that in the long run, it will do us harm – thus going counter to the natural law. (A modern version of a natural commandment that springs from that reality says: "never mess with mother nature....")

All right. We can conclude that there is mounting evidence, from all fields of science, that humans are flawed from birth; let's now explore the possibility that, if we were created by an infinitely wise and loving being, that same creator would see that things were not going well for humanity.

Let's consider the likelihood that the benevolent Big Banger would decide to remedy the flaw in the human condition. It does make sense, after all, to assume that an infinite being would not want to see the highest of the created species – the one most similar to God herself – destroy the planet and themselves.

And there is an additional consideration; it comes from the concept of intergenerational fairness. Is God being fair if she lets my generation suffer due to the sins of the prior one?

What sense does it make, in term of equity, that a flaw caused by prior generations is passed on until it becomes the tragic flaw of a whole species? Why should the willful acts of our ancestors cloud the free will of all future generations, leading us all to a precipice of self-destruction that negates our very existence?

Or, put another way, as in the colloquial expression, why should the "sins of my father" be blamed on me?

Modern psychology offers many examples of how our species is confronted with the reality of having to come to grips with an acquired, pathological trait. It is perhaps most widely applied in the area of alcohol and drug addiction. And it is evident that destructive addictive behavior is often, if not always, connected with clinical depression.

Addiction & Clinical Depression

As stated before, modern psychology characterizes many human frailties (formerly called "sins") as being the result of "vulnerabilities" caused in the development of the psyche, as it is impacted by circumstances. A good example is the person who falls prey to an addiction to gambling, alcohol, or other addictive condition.

A woman is trapped by her own tendency to laziness; she does not take the child to school and feigns an illness. A man is trapped by his own exaggerated sense of sexual conquest and soon starts a very unhealthy relationship with a subordinate at work. Both the laziness addict and the over-sexed workplace abuser lie to cover up their deficient behavior.

Eventually, either the family, or the workplace environment, or both suffer. Children grow up with insufficient education; sex-abuse claims fly in the workplace. Societal resources have to be spent to remedy the damage caused by the narcissistic behavior. Poverty ensues, in greater measure the closer one is to the bad behavior.

Often, a cycle of poverty or sexual abuse ensues. It is a cycle that is so prevalent in human history that some thinkers have despaired and concluded that there is no solution – that the human species is unredeemable.

Addiction to drugs and alcohol is frequently linked to clinical depression. And that is a volatile mixture. When you combine depressive traits with addictive traits, you have reached the ultimate in human dysfunction; inevitably, it is followed by break-up of the family, homelessness, criminal behavior, suicide.

This book is not the place to study what in the United States has become a national epidemic. But the numbers don't lie: As many as 12 million Americans, or about 4%, suffer from clinical depression. That number, coincidentally, is about the same as suffer from alcohol or drug addiction. And it is about the same as are considered to "suffer" from sociopathic behavior.

Notice that I put in Italics only the number that "*suffer*" from sociopathic behavior; the reason is that my amateur instinct (based on considerable experience and reading) tells me that sociopaths, of all mental health "patients," are a type of patient that is morally responsible for their condition. If you analyze all the mental-health pathologies as to cause, giving weighted ratings to nature, nurture or free will, I am confident that sociopaths have the highest volitional component. It is my hope to tackle that matrix analysis in a future work.

Mind, Body and Spirit

In conclusion, *the equation that defines the vulnerabilities or flaws in the human condition must take into account nature, nurture and free will.* How and to what extent each variable interplays with the others is a matter worth exploring, analyzing, and measuring. I suspect it will take many decades, if not centuries, to map the human genome, plus the human mind-psyche, in an integrated fashion that allows some quantification of those three variables.

It is not a simple equation. That is obvious. But it should not be hampered by narrow-minded scientists who seek to eliminate any of the three variables. They are all there: body, mind and spirit. You see them in my high-school banner; and it is the highest-performing high-school (St. Anselm's) in the nation's capital (Washington, D.C.) of the most powerful country in the world.

You see them in the t-shirts of a natural-food store run by my nephews in Miami. You see them in everyday life, and sense them in the lingo of educated people as well as the uneducated. *"Mens sana in corpore sano"* was the Roman slogan; but in the inner-city of America, you better add some notion of God to the healthy body and mind, or you will lose every pastor of every church, and if you lose them (as every politician can tell you) you lose the inner-city.

Body, mind and spirit. You see them everywhere you go – except in the scholarly writings of the anthropologists. And that is a shame, but we don't have time to deal with that.

We must move on to solutions. And in that vein, I humbly submit that Western civilization is built on the premise that redemption is not only possible, but actually happened, and that it entailed a direct intervention by the same Big Banger that sent the universe on its way 14 billion years ago.

How that divine intervention allegedly happened, approximately 20 centuries ago, is just as startling. Not only that, but it is enticing enough to be the stuff of plays and books and movies. It's a great narrative. It's been called the *greatest story ever told*.

But is it convincing? Is it logical? Is it the only possible solution to an admittedly tough problem, which is the reality of a flawed species?

Did Redemption Have to Come in the Form of a God-man?

Based on everything I have said in the prior chapters, it is quite plausible that God felt he/she had to intervene before humans destroyed themselves. But did redemption have to happen by the appearance on this earth of something as unbelievable as a god-man?

The most celebrated Christian theologians are convinced it had to be that way. Here's a typical pronouncement by the most prominent of twentieth century Lutherans, Dietrich Bonhoeffer, as explained by Eric Metaxas:

> *Bonhoeffer was a champion of God's idea of humanity, a humanity that he invented and, by participating in it through the incarnation, that he redeemed.*

Which is certainly inspiring, but not necessarily convincing. The theology is there, but not the empirical science. Remember that in this book, we deal with science and reason first; revealed texts are not authoritative in themselves. The bible is a nice story; but are its teachings compatible with science?

Or put the question this way: Is the Incarnation a necessary component of the Judeo-Christian model of human nature, as understood through the methodology of science?

I have often pondered that. And I have read many a treatise on the concept of the Incarnation and the supposed, rather mysterious way in which God reportedly sent "his only begotten son" to share with us the human condition, including death, so that we could be "freed from death" and saved "unto everlasting life."

The conventional argument or syllogism can be stated as follows. Mankind strayed from the path laid out by God; mankind could not possibly be reunited with God in a flawed condition; therefore, some corrective measure had to be taken by God to eliminate the congenital flaw in mankind.

For the moment, I will skip over the second premise (that man must be redeemed in order to sit with and enjoy the company of God). I will get to it later, when I discuss what might happen if mankind's inherited flaw is not corrected. I want to concentrate, at this point, on the conclusion of the syllogism.

I want to discuss the reason why it may have been necessary (or useful) for God to take the shape of a man and live through so many trials and tribulations. I want to analyze whether that was the only way to reconnect with our species.

The best way to do it is to consider what options, if any, were available to the creator of all things. In my humble opinion, the Incarnation was not the only means available to achieve it. (Though it was certainly the most noble.)

Let's review some of the alternatives that the Big Banger could have used to straighten things out.

Individual Salvation (The Zap-Them-As-They-Die-Model)

Logic or common sense tells us that God could wait until individual humans die and then apply a personal grant of what Christians call "supernatural grace" to those who deserve eternal life.

You are chosen to share supernatural life, to sit at the right hand of the creator, simply because you successfully completed your tour of duty as a member of the human species. You are entitled to an honorable discharge.

This is a very realistic scenario. There is nothing that prevents an infinite God from simply lifting each human species way past the human limitations and bringing each, one at a time, to his/her side.

God can do that. God could program the end of each human life so that the individual simply transcends a human form and acquires some form of eternal existence.

Yet, for whatever reason, it is apparent that the Big Banger did not want to do it that neatly, that precisely, that individually. I would guess that the creator of all things had a more sophisticated, more collective, more holistic model of remediation.

But in theory, at least, there are other models.

Instantaneous, Collective Salvation (The Start-From-Scratch Model)

An alternative model is one in which the Almighty would realize that the entire initial plan simply did not work out. God could easily decide that giving our species the right to choose wrong over right, selfishness over sharing, pride over humility, just didn't work out. God could decide to turn all of us into animals, who follow carefully calibrated instincts, eat or kill each other as needed, die without crying, and end the natural life cycle without much fanfare.

He/she could have just turned off the knob that allows humans to choose, effectively turning all of humanity into smart apes. This is not idle talk. God, by definition, has infinite power, and that includes the power to start from scratch.

Non-intervention Model (Let-Them-Destroy-Themselves)

One can also conjecture that God could have allowed humans to simply destroy themselves.

Why is this scenario unlikely? Only for one reason: because it doesn't fit with historical facts. Theoretically, it is as likely as any other. God creates man. Man disobeys God and messes up the entire system, such that it becomes self-destructive. God takes a look at it and makes a decision: the human species is just not worth the effort.

Let them die, kill each other, destroy their planet. Humans, God could conclude, are not worth saving. It is entirely plausible. Entirely rational and predictable. But not what actually happened, as far as we can tell from the evidence.

What actually happened is that God decided to stick his/her hand into history. God, as far as we can figure out, chose to push the envelope, to do something akin to what humans do, when we realize that if we "can't lick them, join them."

God chose to join humans so that he/she could lick the ails that affected them. And this decision is the one most consistent with the idea that God is love.

The He-So-Loved-the-World Model

As with creation itself, the best explanation lies in the idea of love.

A great philosopher once said that "we cannot love what we cannot imagine." I think that is self-evident. In my humble opinion, the Jesus story is comprehensible only in that context. Otherwise it makes no sense.

As suggested above, the creator of all things probably did not need to assume the shape (and mind-psyche) of *homo sapiens* in order to correct the flaw that had developed in the species, by early humans. God could easily have restarted his/her creation of humanity by (1) zapping all living humans with a reordering dose of supernatural personality-adjustment, or (2) by starting the process all over again with a new set of ancestral *homo sapiens*.

Once again, I was jolted out of the above rumination by a passage from Metaxas's biography of Bonhoeffer. To think that God could simply wave her magic wand and redeem humankind is to think small; and God is much, much bigger. Explains Metaxas

> *Bonhoeffer was thinking in a new way about what he had been thinking and saying for two decades. God was bigger than everybody imagined, and he wanted more of his followers and more of the world than was given him.*

Bonhoeffer saw God as a jilted lover, who wants more reciprocity. In effect, he was saying that God's love is not an abstract concept. God loves each and every one of his/her creatures, including the ones who came before Jesus and the ones to be born later. God chose the most complete, efficient, and exemplary means to make humanity whole.

"Exemplary" is the key here. God's decision to assume the figure and substance of a human was to teach us, by example, how to love. And to do it in the flesh.

Imagine a believer trying to follow God's mandates, obey God's will no matter the circumstances and the pressures of a nature that craves adulation, comfort, and leisure. What example would such a believer follow?

I suppose, if they knew their biblical history, that they could imitate the prophets or St. Augustine or St. Thomas Aquinas. But the prophets lived in a totally different era and we know little about them, except heroic acts and special revelations (allegedly made by God through angels and burning bushes and whatnot). And Augustine/Aquinas lived in a different era, subject to totally different allures and impediments to earthly welfare.

Neither the prophets nor the saints are particularly useful models to follow.

Jesus was a regular guy. He was born into a lower middle-class family. His people were not considered high-class. They were subjects, not citizens. Jesus overcame his roots, and did so with flair.

For me personally, the life of Jesus is uplifting in so many ways that I could write a book about it. (In Chapter VI, I give some examples of lessons from the biblical story of this amazing human being, whose words and deeds were so well calibrated, so inspirational, so uniquely charismatic that they have no equal in history.)

Jesus set the best example one could set: that of a decent, working guy who tries to do good and ends up being crucified for it. One thing is to kill the messenger when he brings bad news; but why kill the messenger when he brings a Gospel, which literally means "good news"?

Jesus was the classic case of one day being the toast of the town and the next day being rejected by your friends and family, for absolutely no valid reason. That example should help all of us who read about it and choose to believe it. It happened to me in a magnitude not quite as extreme as the Jesus story, but quite comparable in the rather unexplainable, undeserved change of fate.

I share it with my readers as an example of how every one of us can benefit from knowing that at some point, each and every one of us must accept some amount of injustice with some measure of forgiveness.

My Short Life as a Rejected Prophet

In 1997, I was elected the first "executive" mayor of Miami. I had previously served three terms as mayor, but in a totally "legislative" capacity: the city manager, appointed collectively by all five city commissioners (the mayor being one of the five), had all the executive authority. He appointed the police chief, the department heads, and all other employees, subject only to civil service rules.

That had all changed with a reform approved by the voters. In 1997, I had been certified as the winner in a hotly contested election; I proceeded to appoint a new city manager, with clear instructions that he should "clean house." Henceforth, Miami's upper-level bureaucrats would no longer be the pampered, overpaid class that spends most of their time complicating the lives of the citizenry.

For my efforts to streamline the bureaucracy, I faced the most unfair attacks ever unleashed against a municipal leader. The print media, in particular, went after me in a vicious way. They said that my election was fraudulent; that my efforts to clean house were less than honorable; that I was guilty of association with bad people.

Every single charge leveled against me ultimately turned out to be false; the electoral fraud was concocted by law enforcement in what is called a "sting operation." Neither I nor my family or volunteer campaign workers had done anything wrong.

The police chief that I tried to replace was subsequently convicted of grand theft; his co-conspirator, who cooked the books as accountant for two trusts managed by the chief, committed suicide, but only after leaving a note that explained all the fraud committed.

The pressure from the media, which used their investigative resources to the maximum, ultimately caused the courts to invalidate the election and install my opponent in the mayoralty. This fellow, who had long before been dubbed "crazy Joe," ultimately spent a night in jail for hitting his wife with a teacup container.

The print media, and in particular one editorialist, conveyed to their readers the notion that I was mentally unstable – that my well-crafted analyses of city finances were delusional and that I should be institutionalized.

It took me almost two decades to recover my reputation and a measure of political power. The same newspaper that made me sound like a lunatic now supports my candidacy for county commissioner and lavishes praise on my initiatives to solve the county's ills.

In a metropolis of almost three million people, I have the support of not only the mainstream media, but just about every major social media commentator. My popularity with both the intelligentsia and the people at large is higher than ever.

How I overcame the media attacks is a longer biographical project. What is of interest here is the inspiration I drew from the life of Jesus, Gandhi, Martin Luther King, Lincoln, and Winston Churchill. All of them endured periods in which they were attacked by the "establishment" as unstable, over-ambitious demagogues. They are great examples to me.

But the personality of Jesus has a unique transparency of motive that resonates more than even those heroes of history. As told to us by his followers, Jesus had the power to destroy his detractors; he could ostensibly "rebuild the temple in three days," presumably resurrect the dead; he could bring hell and fire on those who might foolishly attempt to torture and kill him.

But he passively accepted the humiliation of having the populace choose a proven, heinous criminal (Barabbas) over him. Humility was imposed on me by circumstances that I could not avoid; for Jesus, it was self-imposed. Why? Because he wanted all of humanity to learn the limits of love.

In choosing to live the life, and suffer the worst indignities that can befall a human being, Jesus (If he was God) clearly wanted to leave an example of the potential that humans have to care for others. *He was not a perfect being; he was a perfect human being.*

The aftermath of his life is the most telling thing of all. Jesus changed the course of history. There is no other prophet, in any other religious tradition, that left such an imprint on history.

What Other Religious Traditions Say

At this point, I have to state what we, lawyers, call a "disclaimer." It is not the only one in the book. There will be other moments – particularly in my treatment of history (Chapter V), where I disclaim being anything close to an accomplished historian.

Both the history and the religions of Asia are blurry to me. I haven't read enough and haven't formally studied the histories of China or Southeast Asia. India and North Africa are in a different category – the former because of my extensive readings by and about Winston Churchill.

As for North Africa, being part of the Mediterranean basin and of the Roman empire, the region falls into what we generally regard as the West. Like the Middle East and India, they can be lumped in with the West in our analysis.

Sub-Saharan Africa is a whole different matter. Except for the anthropology of the first pre-humans, including *Australopithecus Afarensis* and *Australopithecus Africanus*, who lived millions of years ago, I am equally ignorant of their history and culture.

But I have picked up enough, from more learned authors, to know that the idea of a God-man, or a specially anointed Messiah sent by God to redeem mankind, is not limited to Jewish and Christian tradition.

I offer my reader a quote on that topic; it is taken from a book by Karl Stern, titled *The Third Revolution*. (Stern, I should add, is best known for his written exchanges with Carl Jung, from whom he learned psychoanalysis. He also had a lively line of communication with Dorothy Day, Jacques Maritain and Teilhard de Chardin.)

Here's Stern on the Incarnation:

> *When Saint Paul speaks of 'God who at sundry times and in diverse manners, spoke in times past to the Fathers by the prophets,' many hold that he does not hold to the British tradition alone. The prefiguration of the incarnation can be traced in all people; in the Hebrew people it can be traced in a special way.*

I accept, for purposes of discussion, Stern's view that the incarnation of a god-man has global cultural roots. It is not all that important to the discussion, or the conclusions, of this book.

What is important is to analyze its meaning within the real-life Jesus story and what Christians consider its value as a "redemptive" event. Redemption, understood as reconnection of humankind to a creator, is one thing. The idea that such a reconnection required for the creator to take the form of one of his creatures is another thing entirely.

Later, when we discuss the figure of Jesus, I will renew the argument made before which holds that the Incarnation was simply the most effective, most impacting and most noble way of "rebooting" the flawed human computer by reconnecting it to the perfectly wired, infinitely powerful divine computer.

And yet another question beckons us. It is the interdependence of humans with their own kind. They say "it takes a village to raise a child." Perhaps, but why does it take a village to reconnect me with God?

Why can't I do it alone – assuming I want to do it in the first place?

To that interesting question I dedicate the next chapter.

Chapter V

MAN: THE SOCIAL ANIMAL

The doctrine of the Trinity reconciles the paradoxical affirmations that God is self-sufficient and that he is love. H.P. Owen

God hath thus ordered it, that we may learn to bear one another's burdens; for no man is without fault, no man without his burden, no man sufficient of himself, no man wise enough of himself; but we ought to bear with one another, comfort one another, help, instruct, and admonish one another. Thomas à Kempis, *The Imitation of Christ*

At first blush, it does not seem fair, or logical, that man should have to rely on the good faith of others to shape his destiny. An omnipotent creator chooses to give one species free will, and to hinge eternal life on the proper use of that faculty, yet makes the entire process subject to the choices made by legions of people who roam in the vicinity of our lives.

It would seem more logical that our species would interact with the creator individually. That is the way of modern technology. By way of example, I can deal with my 5,000 Facebook friends individually, if I want. Why, then can God not deal with me individually? Why must

I rely on my brother, or my neighbor, or my wife or my kids (or even grandkids) to decide my fate for all eternity?

One thing is clear on this. God could have created an intelligent being, with free will, and with a body like ours, that would not have to interact with other beings, also possessed of a free will. He/she could have created us as single-gender, self-sufficient entities who need no one but ourselves.

In short, God could have created beings who have a strictly *vertical* relationship with their maker. Instead, it is clear that there is a *horizontal* component to the equation by which humans interact with the presumed Big Banger. Humans relate to God through each other; for many, that is the only way in which they experience spirituality – by which I mean the dimension of existence that goes beyond (transcends) the more basic effort to satisfy survival needs.

Man is more than the sum of his individual needs, more than the satisfaction of his appetites. Man is a social animal, with an urgent need to relate, to attract, to interact and to love other humans. The French philosopher, Jacques Maritain, put it thus:

> *By the very fact that each of us is a person and expresses himself to himself, each of us requires communication with the other in the order of knowledge and love....By its very essence, personality requires a dialogue in which souls really communicate.*

Maritain was right. Human nature requires interaction, communication, love. Some of the time, the effort to interact with other humans begets disastrous results for the individual. Man knows that, and yet man (and woman) persists in the effort to love and be loved.

That is evident. Humans are defined by their love affairs. Some philosophers and psychologists subdivide the general concept of love into categories such as *caritas* (charity), which seeks nothing in return, *eros* (or sexual attraction) and filial (brotherly and sisterly) love. Plato argued that the last one of those can only be had between perfect equals, who don't need anything tangible or concrete from the other. Just friendship, camaraderie, companionship.

However one ranks human relationships, it is clear that our species enjoy each other's company. That much is evident. The question is: Why did God make humans social beings? Is it a reflection of his/her own being?

Is God a communal being? Is he/she perhaps triune, as Christians believe?

Why a Communal God?

Christian theologians are convinced that the creator is not a single, monolithic entity. They would have us believe that God not only took the form of a man, but that the man (Jesus) was not just a manifestation of God, the creator (previously referred to as the Big Banger), but a separate "person." Furthermore, they argue, the creator (which they characterize as the "father") relates to the God-man (characterized as the "son") in a way that reflects a third reality – a third person, if you will.

I have spent a lot of my leisure time reflecting on this, applying my reason and everything I know about science. I have done it both deductively and inductively. By that I mean that I have researched it both by pondering inwardly towards the core of my being and by analyzing the findings of the greatest scientists in the various different fields – from biology to anthropology, from physics to astronomy, from genetics to demography, from economics to history and sociology.

In this chapter, I share with my readers what I have concluded, no-holds-barred. My conclusion is that there is a likelihood that God, if he/she exists, is a community of persons or entities. This conclusion is impossible to prove, using the methodology of natural theology.

There is no way to prove empirically the triune nature of God.

But that doesn't mean that we cannot argue it is a reasonable – maybe even a probable - conclusion. In that sense, it is no less scientific than the notion that the universe began at a singularity of infinite density (as indicated by the equations of relativity) or that in the next few fractions of a second, it stretched into a fabric whose stitches are defined by a quantum of a second and a Planck unit of distance.

We can't prove, let alone measure, any of those phenomena, yet they are reasonable (and likely) discoveries of human reason. They are consistent with other facts which can be measured, as well as common sense.

The only real difference, in terms of the scientific method that leads to each conclusion, is that one scientific reality originated from a human hypothesis and the other from what is allegedly a "revealed" text. There is no likelihood that reason alone would ever have prompted anyone to suggest that what appears as an infinite force, of infinite dimensions and intelligence, has three components.

But it is not far-fetched to think it entirely possible, as we shall see.

God is infinite. I say "by definition" because of the logic of the succession of events that led to the Big Bang, which was a moment of infinite density, when huge amounts of matter all of a sudden appeared, sitting astride a non-dimensional space in a previously non-existent period of time. That syllogism was already discussed in the first chapter.

Let's restate that. That's not my imagination or my "faith." That's what empirical science teaches us. Let's take a couple of minutes to parse it out, in the simplest possible terms.

All that we see; all that we measure; all that we sense apparently began 13.7 billion years ago, at the Big Bang. As far as we know, there was nothing physical before that moment. The only thing that existed is some sort of timeless ("eternal") being that had the power to create everything that was propelled at the Big Bang, along with the magnificent order that was apparently instilled in a quantum of a second in that mini-universe.

We know that the Big Banger had to be there right before the Big Bang, otherwise there is no one and nothing to bang the drums of the early universe. But was the Big Banger *always* there?

In this analysis, we must resort to pure logic, because science cannot give us fool-proof answers. This kind of analysis is what scientists call "metaphysics." Lay people prefer the term "common sense," because it is a commonly accepted idea, passed down from generation to generation pretty much among all civilizations.

"Nature's God"

The idea of a deity reflects a consensus that can be gleaned from pretty much all cultures, including the non-religious deists, such as Benjamin Franklin. He and many of his colleagues who founded the American nation – and who were not particularly "religious" in the sense of going to church – referred to the Big Banger as "nature's God." To them, as well as to almost everyone who has looked at the matter dispassionately, it is reasonable to conclude that the physical world was created by someone or something that had infinite power and an infinite prior existence.

Otherwise, he/she too would have had to be created.

Fair enough; and to carry on the logic, in a symmetrical way, I suppose it makes sense that what is infinite in one direction of time is infinite in the other. (Without getting too technical, science is about as clear as can be on the issue of what is called "the arrow of time." For scientists, the movement of time is nothing more and nothing less than increasing entropy, i.e., decreasing order. Since there is no indication that God is changing in any way, then God must be eternal.)

Not surprisingly, humans find it easier to embrace the idea of an infinite future than an infinite past. There is a reason for that. Most of us, most of the time, don't want to die. Like the song in the movie "Fame," we "want to live forever."

Not that we can fully understand what that means. We live in time; we live in a confined space; we have limitations (in terms of speed, the speed of light; in terms of space, the close vicinity to our planet and its moon; in terms of duration, a life not exceeding 114 years for men and 116 for women; in terms of intelligence, an IQ somewhere below 180; in terms of size, about seven and a half feet.)

What Sense a Trinity?

The yearning to live forever has been discussed already. We humans acknowledge that we are not eternal going back in time, but we (foolishly perhaps) convince ourselves that we could be eternal going forward in time.

We presume that we could be adopted by the infinite being that created us and who will keep us for as long as the infinite being wants us around.

In one sense, it is a bit of wishful thinking – the height of self-conceit. Our material condition limits us in speed, duration, size and intelligence, yet we think we can live forever, if we can only please our creator.

Let's accept, for the purposes of discussion (and since we have no proof to the contrary) that particular postulate of religion – the one that suggests a possibility of life after death. But to what end? Once I am dead, I cannot improve the lives of others; I cannot hold and hug my wife and kids. I cannot write books to enlighten the lives of my descendants. I cannot invent a better mousetrap or ride a nicer car, or taste a better brand of ice cream.

And, with all due respect to Islam, I cannot really look forward to having many virgins for my enjoyment. Somehow, the idea – while possessed of a certain allure – does not stand the test of fairness, since I cannot imagine the young virgins wanting to spend their heaven in my company. Plus, I would rather imagine being in love with one woman (my wife of 40 years and counting) and living in a place where that love grows with time and ever-increasing joy.

I like, in that sense, the C. S. Lewis notion of heavenly existence as an eternal dance: a never-ending party where beauty and music mix with discoveries of science and insights into the soul of my kids and my friends; where friendships strengthen and enmities fade; where the broadcast news, as it reaches us from earth and from heaven itself, is increasingly better and better.

Sir Thomas More called it Utopia, when positing a heaven here on earth. Others call it paradise. The standard word is heaven; and it is also applied to the sky, under the assumption that heaven is located somewhere up there, past the clouds that we see.

Fair enough; but does paradise require a triune God? Does the human psyche yearn for a three-person God? Is there scientific sense to that theological invention? Or is it just like the proverbial question of "how many angels can dance on the head of a pin?"

St. Augustine Tackles the Trinity

The great philosophers – and in particular the Christian members of the A-team (Augustine and Aquinas) spent a lot of their time on the idea of a triune God. It is said, in the case of Augustine, that he was walking on a beach, meditating on the topic when he saw a child making a hole in the sand. He asked him what he was doing and the child replied: "I am taking all the water from the ocean and putting it into this hole that I am making." To which Augustine reportedly replied: "That is impossible; the hole is too small and the ocean is too large to fit in there."

The child allegedly scored a major oratorical victory over the great philosopher/theologian, when he ended the exchange with a flourish, saying: "It is no harder for me to put all the ocean in this little hole than for you to put the enormity of the trinity in your little head."

Taking a cue from that fable, and injecting a dose of modesty into my little head, I will step back from trying to understand a trinity and work first on trying to understand whether God is, at least, binary.

I refer to the two genders.

Keep in mind, as we delve into this presumably less tricky bit of theological quicksand, that psychology is a modern science and its precepts, except insofar as they were purely intuitive, were not formulated at the time of Aristotle, Augustine or Aquinas.

In those times, there was no understanding of the compound composition of human genders. I expect all three savants would have been shocked that science today sees male and female components in all human species, and that the person's psychological gender is defined by the preponderance of male traits in the man and female traits in the woman.

We know better; and based on the premise that humans are made in the image and likeness of God, we intuit that God is both male and female. Our intuition is borne out by contemporary thinkers.

Is God Male, Female or Both?

One source to consider when analyzing this issue is the biblical Jesus.

If, for whatever reason, God assumed a human form for about 33 years, that person had to have some very impressive traits. We can't even begin to prove it, but we can derive insights from the story, confirm that they are consistent with science, and see if they help in the analysis.

Let's assume for the moment that Jesus was the human face of God. What was he like? Was he an alpha-male, as depicted in most paintings? Was he what modern writers call "metro-sexual," meaning the sensitive, soft-shelled, high-pitched voiced, caring character from the play *Godspell*?

Let's see what the academics say.

In a scholarly lecture presented at Saint Mary's College in Notre Dame, Indiana, Sandra Schneiders performs an in-depth survey of Gospel texts and concludes that the Gospel "portrays Jesus as non-aggressive, non-competitive, meek and humble of heart, a nurturer of the weak and a friend of the outcast." To Schneiders, these traits place Jesus more as a female than a male archetype. She expands:

> *He [Jesus] manifested and preached a spirituality that was characterized by stereotypically feminine rather than masculine enough for men to assimilate, but it has always supported the spirituality of women.*

I myself, as a lawyer and politician (and, I might add, as a grandfather of eight), am drawn to the metro-sexual personality as much as I am drawn to the alpha-male one. It's kind of a simple dichotomy: *The former works in times of peace; the latter in times of war.*

Most of the time, up to now, and hopefully all the time in the future, we should yearn to be at peace. Most of the time, we should be drawn to the sensitive, baby-kissing, ultra-forgiving, non-judgmental side of Jesus.

This was the side of Jesus emphasized by Karl Stern in his marvelous book, *The Flight from Woman*, in which he dissected the personalities of some of the great philosophers and showed that those who disconnected from their feminine side were incomplete and borderline dysfunctional.

Agreed, but then why did God take the form of a man?

I have always assumed that Jesus, as a representative of God – or even the "son" of God – appeared among us as a man because, in those times it would have been logistically impossible to appear as a woman. Certainly, it cannot be denied that Jesus was a biological man. He was probably, as we shall see, masculine to the core in voice, stride and overall physique.

Interestingly, some passages in the Gospel refer to him as the "son of man" and some as the "son of God." Christian theologians have interpreted that to mean that there is a relationship between Jesus and his "father" in heaven that bespeaks of that of an offspring. The same theologians vehemently argue that the divine side of Jesus was not created 2,000 years ago, but "begotten" at the edge of infinity, where time does not exist. They are sure that Jesus is the second "person" of a "trinity" that includes a "father," a "son" and a "holy spirit."

That sort of conjecture is way beyond the scope of this little book. My focus here is to determine if God has more than one person – not how it came about. I leave that question to much greater thinkers.

Back to the Trinity

I ask my reader to ponder for a second two ideas, and only two ideas. Assume two premises, if you will.

Premise no. 1 is that God, as an infinite being, has no limitations. There is nothing greater than God; nothing more interesting than God; nothing more complex than God; nothing more dynamic than God.

Think about that for a second. God, by definition, must be interesting, complex, dynamic.

When we meet a person who is stuck on one subject – let's say a very serious, dogmatic, religious zealot, we often find him tedious and uninspiring. When we meet a person who is diversified, eclectic, equally amused and knowledgeable in the arts as in the sciences, able to relate to kids as well as adults, capable of discussing serious subjects but also capable of telling the best jokes, comfortable in the company of athletes as in the company of out-of-shape, absent-minded "nerds," cool to anger but hot to passion on subjects that merit passion, dogmatic on the fundamental rights of man and flexible on the aesthetic preferences that people choose as they exercise those rights – when we meet *that* person, we are sold.

Now hold that thought.

Here's the second thought. What makes life interesting is the interaction among humans. We are not built to live alone.

I am an authority on the subject of being alone, and liking it. I love being alone, in the sense that (1) I like my privacy, so that I can dress the way I want, do what I want, rest or move around if I want; and (2) I don't want to constantly interact with those who happen to be in my proximity.

In that sense, I am a loner. An introvert.

But that aloneness, that self-contained, self-sufficient trait that is emblematic of my personality goes only so far. If I had to live totally alone, even with all the creature comforts known to man (food and shelter aplenty), and even with tons of good books and movies to watch, I would soon wither and die of sadness.

The only difference between me and my wife in this matter is quantitative. Whereas I could not resist to live alone for more than a week or so, before getting depressed, my wife could not resist to be alone for more than an hour or so, before getting depressed.

I am not kidding about that. We are that different. Right now, she is among roughly fifty or more friends and family down at the beach, at the water's edge. She is as happy as a human being can be, surrounded by kids and adults of three generations and loving every minute of it. I, on the other hand, am happy as a lark, sitting by myself writing this book, subsisting only with a bit of rum and coke! We are both doing what we love; but soon, in a couple of hours, I will be yearning for her touch, her voice, and for the reaction of those I can copy with this text.

I cannot be happy by myself. I cannot be complete without the realization that my ideas will influence others of my species. I cannot be forever isolated, among books and films and past histories of human activity. I want to share in the human condition. Perhaps not as intimately and certainly not as shoulder-to-shoulder as my wife.

But I cannot live alone. Apparently, neither can God.

The Deity as a Community

God is too great to be a singularity – a single, definable reality that has infinite dimensions. It doesn't help the description of God to say that he/she is infinite. Infinite is still a static concept. In math, it is represented by a number eight standing on its side.

That doesn't cut it. That is not interesting – let alone inspiring.

If God were just one person, even an infinite person in time and space, God still wouldn't be the "most interesting man in the world" (as in the beer commercials), let alone the most interesting woman in the world. God is multi-faceted, multi-talented, multi-diverse, multi-gender and multi-person. God is the "most interesting thing in the universe."

I kept thinking about that description, and a song came into my repertoire of ideas. It is perhaps fitting that only a song can define God. And it is fitting that is the song that best describes love.

God is a "many-splendored thing," just like love is a "many-splendored thing."

Again, this is not a leap of faith, but a leap of logic. If God exists, if God is the explanation for the Big Bang, and the Inflationary Moment, and biogenesis (for all three of which there is no other scientific explanation), God is most likely a many-splendored reality.

Intuitively, we could conclude that God, if he/she/it exists, is probably an interacting community of entities (we call them "persons") that is dynamic, loving and eternal.

This is not to suggest that God must be *exactly three persons*. Thinking as a holistic, secular scientist (or, as previously defined, as a "natural theologian"), I would argue that pinning God down to three persons is a simplification concocted by theologians, based on biblical insights.

And yet, it makes all kinds of sense, though the argument has to be made – as metaphysical arguments often are – by using metaphors. The one I prefer is called a "hologram," which is a device similar to three-dimensional computer graphics.

The metaphor of the hologram derives additional power from the fact that it provides an analogy to our human reality. In the limited human lens, God appears as a trinity of persons.

The reader might find another useful metaphor in the idea that humanity is inside a submarine and is using a periscope to try to understand the complicated weather above the surface. Viewed by the human periscope, sticking out of our submarine and peering into the apparently infinite vastness of the cosmos, God appears as three persons.

The submarine metaphor is a modern version of Plato's cave, in which humans can't see objects directly, but must rely on the shadows created by the light that penetrates from the mouth of the cave. The metaphor of the hologram is all mine, although it was probably inspired by something I read or heard.

What is a Hologram

A hologram is a machine that projects a three-dimensional picture onto a two-dimensional screen, letting us see an object in a way that cannot be seen when looking at a picture. It is our best shot of seeing three-dimensional reality in two dimensions. It works by concentrating a laser beam of light onto a screen, after it impacts a solid such that it captures views that cannot be captured on regular film.

A hologram, in other words, is a three-dimensional picture. It represents living images as seen by a light more powerful than our eyes, but technically able to show us a view that is more complete and much more accurate than a camera can obtain.

God has infinite dimensions. God is not limited by space or time. But *if we could view God* using a concentrated laser beam of light that captures God from all angles – if we could view God through a hologram – God would appear to us as a three-person reality.

Why do I think it is three persons? Why not many more? Again, as I said in the preceding section, I am not in the business of limiting the scope, or the traits, of an ostensibly infinite being.

Nor am I enthralled by the number three.

Two very smart people who read my writings react with skepticism when I mention the three dimensions of the physical universe.

One is an MIT-trained engineer and another is a Villanova classmate who had, by the time he entered college, read pretty much all there is to read in the liberal arts. You can say one was the ultimate technocrat and the other was the ultimate Renaissance man. (If you could fuse them together, you would get a twenty-first century polymath.)

Those two great minds are not alone in questioning the scientific logic of a trinity, or the related idea that one of the members of that trinity was both God and man. The great Isaac Newton, who was a firm believer in God, was not so enthralled with the concept of a God-man.

Newton questioned the notion of "consubstantiality," meaning the idea that Jesus had both a human and a divine "substance."

Newton v. The Trinity

Here's how Newton biographer Mitch Stokes explains the Newtonian dilemma:

> *In addition to believing that consubstantiality was not a scriptural doctrine, Newton believed that the metaphysics underlying it "is unintelligible. 'Twas not understood in the Council of [Nicea]...nor ever since." Just what is a substance, and what does it mean to be of the same substance (and not merely the same kind of substance)? "Substance" is a philosophical term that is mysterious at best. Like Locke, Newton believed that, even if things possessed some underlying substance, we know little, if anything, about it. And if this is so for ordinary material objects, how much more in the case of God?*

In one sense, I agree with Newton. Given that we have a hard time understanding the essence of ordinary subjects, how can we claim to understand the essence of God?

I want my reader to know that I myself have my doubts about this notion of a trinity. Why does God have to be thought of as three? Why not just one, or an infinity of dimensions?

As I stated of the method I use, it is fundamentally scientific; and while it does not ignore, let alone reject, insights that might flow from allegedly "revealed" texts, it does not grant them any sort of intellectual immunity. My method does not accept any notion of "infallibility" by papal pronouncements or the bible itself.

This methodological constraint bears repeating: Revealed texts are considered helpful to the analysis, but are by no means the essential building blocks in the analysis.

Yet I have to admit that the three-dimensional nature of reality constantly rears its not-so-ugly head. Where we would love to see one-dimensional simplicity, or two-dimensional creative tension (of the kind that hopefully exists in a marriage or a two-party electoral system), we keep finding three substances or realities.

Let's take the alternative of dual substances, since I have already devoted much of my argument to debunking the idea of a single reality (matter) in prior chapters.

Pure materialism doesn't hold water, but there is a certain allure about duality. Things do seem to have an "off-on" character. Positive and negative. Ying and Yang. Male and female. High and low. Thesis and antithesis. Liberal and conservative. Young and old. Physical and intellectual. Mind and matter.

But duality seems to be an incomplete equation – almost as if it were a problem not solved. Whenever we meet a classic built-in duality, it seems that the two parts are in tension and that the tension is inevitably resolved when the two extremes fuse together to create a three-dimensional whole. Nature thrives on a universal dynamic, under which antagonistic things seem to be in constant, unifying flux.

Positive-charged particle combines with negatively charged particle and both fuse with a neutral particle. That is the atom and without it there is no solar system, no earth, no us.

Male bonds with female and begets an offspring, forming a family. Ideas have that triune quality: philosophers have labeled them the thesis, the antithesis and the synthesis. The triune reality is particularly evident in the domain of ethics.

Instinct + Inner Voice = Virtue/Maturity

In the classic ethical framework (discussed at length in Chapter I), there is a constant tension between the person's appetites and his/her moral barometer, which tension resolves itself in the decision to act in a particular way.

I should add that often the decision can be an omission – a conscious choice to do nothing. And here I want to digress and engage in some introspection, which I have previously explained is what scientists call "deductive reasoning."

As I sat down to write this section (which I have been pondering for a few weeks before today), I considered taking a nap. The mental process by which I ultimately decided to make the effort to write these lines went more or less as follows.

I was sitting in the balcony, and it was quite breezy and warm, as only Miami in October can be. The book I was reading was serious, but also light in its almost familial recounting of anecdotes dealing with the removal of Cherokee Nation from Georgia and Alabama in the early decades of the nineteenth century. If I kept on reading, I knew I would fall into a very pleasant slumber.

I had no qualms about taking a nap. It was a lazy Saturday, with nothing pressing on my calendar – either from immediate or long-term projects. Yet I ultimately decided to take up my laptop and start writing. My inner voice told me that I might not have another opportunity to put in writing what I had been thinking now for a couple of months. The ideas percolating in my mind could disappear and be lost forever.

There is no immediate reward for making that decision. This book might never be accepted by a major publisher. I certainly wasn't going to get rich from any insights I may have on a topic as esoteric as the divine trinity.

I really could not think of one person in the whole world who would for sure, with absolute certainty, read this chapter of my book. Nevertheless, I took up the (electronic) pen, put away somnolent inclinations, and forced myself to think coherently and compose. The result is this section; the reader can soon determine if my insights are worth the effort it took to write them.

In the final analysis, the decision to make the effort to write down these ideas was based on the notion that I have some sort of obligation to preserve (for posterity perhaps, or for a small contemporary readership) ideas that might be unique. In other words, I convinced myself that the decision to write about my notion of a human equivalent of the trinity is not really mine alone to make.

I have to consider my inner-voice.

In the process, I achieved an intellectual balance – like a compromise between instinct and obligation. You can call it a little bit of wisdom.

Human Wisdom: A Touch of Godliness

We use the adjective a lot more than the noun – except when giving names to our female children, whom we often call "Sophia," which is a noun, meaning "wisdom." For example, we constantly say "he or she wised up" to describe a youngster who realized the error of his ways. Most frequently we use the term to describe a "wise decision," as opposed to a foolish or immature one. Sometimes we use it derisively, as in "what are you – a wise guy"?

We all know what it means to be wise, or unwise. But we don't reflect much on what the underlying parameter is that puts something in the realm of wisdom, rather than foolishness or immaturity.

In essence, wisdom is the fusion of our appetites with our moral barometer. Wisdom is the synergistic result of analyzing our instincts, moderating our passions, tempering our temper and, without losing the sweet, inner joy that fuels our actions, achieve at least a modicum of greatness.

Wisdom, in short, is what happens when we follow our better angels.

Except that (scientifically speaking) there are no angels. And no demons. There is a flawed, fallen, contaminated human nature, that struggles inside a tunnel full of trials and tribulations, whose survivors are those who see the light at the end of Plato's cave and understand that the light can only be reached in the company of other lost souls.

Wisdom is knowing both our limitations and our obligations. In that sense, the Christian ideal of a saint, and of the "communion of saints" is quite compatible with what modern science teaches about mental health. But that is a topic for another book, by someone better versed in the variations of human, mental pathology.

Here I want to continue doing my best to explain how we humans might resemble a divinity that has more than one dimension. Let's analyze our own species and see what we derive from togetherness. (In group therapy, and in my own family, the emotional dimension of togetherness is demonstrated by a "group hug.")

Using that metaphor, we can conflate it with the idea that God is a dynamic community of persons, rather than a single, static entity. I would hate to think that God is not capable of a group hug....

But back to us: Why do humans yearn for community?

What People Do When They Are Together

We start by noticing what people do when they are together – as opposed to when they are alone. Some human functions can only be performed alone; among those are bathing and discarding food and drink that is no longer useful to our bodies.

Eating can be done in private; so can drinking or listening to music. Observing nature and rejoicing in its wonders can be done alone. The same is true of discovery, whether by research or measuring or reading or calculating solutions to equations.

But many of these functions are more fruitful when done in the company of others who share our appreciation for good music or food or a beautiful sunset. Scientific discovery is almost inevitably a communal affair. I am not aware of a simple great invention (from electricity to the laws of motion to algebra or the discovery of the Big Bang or the DNA) that was the result of one person working alone.

(Anyone who thinks that Einstein formulated special relativity by himself should look up what the physicist George Francis FitzGerald wrote much before, containing the principal elements of the relationship between space, time and the speed of light; anyone who thinks that Crick and Watson developed the double-helix DNA without help should look up the woman who never got credit for the idea, named Rosalind Frank.)

Humans become greater, happier, more productive when collaborating with others. Sometimes, as in reproduction, "it takes two to tango." Sometimes, as in raising a child, it takes a village.

The desirability of collective action, of having at least two people engaged in a common pursuit, is evident when the issue at hand is romance. In that context, it is important to note a key component of romantic love: the way human genders complement each other.

And it is hard to argue with the notion that the human pursuit of happiness begins and ends with love.

Is Love the Answer?

Many Christian wedding ceremonies include the passage from St. Paul's letter to the Corinthians. My readers probably remember well how St. Paul (clearly the most prolific chronicler of the lives and times of Jesus) rails about the immense importance of love, by comparing it to other virtues. It matters not whether you have "the faith to move mountains," and "the gift of prophecy," and convictions so strong as to let "your body be burned." Those things are "like a clanging cymbal" if they are not motivated by love.

At this point, we need to go back to the quote that begins this chapter. As per the quote from H.P. Owen, the doctrine of the Trinity "reconciles the paradoxical affirmation that God is self-sufficient and that God is love."

I had no notion of the existence of H.P. Owen when I started writing this book. But the idea that he had was identical to my idea, and it is very simple.

If God is love, how can God also be self-sufficient? Or, put it this way: If God is infinite, and needs no one to empower him/her, or to create him/her, or to interact with him/her, how can God love? What is there for God to love?

The most likely conclusion, for me as a natural theologian, is that God is an infinite being whose interaction with humankind is reflected, in our minds, as a triune reality. It is not a scientifically provable proposition. But it is certainly consistent with everything I know of science.

So what does a loving God do for his/her creatures who have gone astray? How does God bring back humankind without taking away the single, most distinctive trait with which humans are endowed – the freedom to "mess with mother nature?"

The story of humankind's reconnection with God is remarkable not only for its socio-political shock waves, as the news of redemption spread, but for its exquisite timing. What I mean by that is something quite startling: creation could have happened any time. *Redemption had to happen when it did, because otherwise a critical mass of evil would have destroyed humanity.*

This is my own, exclusive, unproven and unprovable theory. It may or may not convince you, but it will make you think and ponder on real events, whose succession happened in a sequence that reminds one of a high-wire act by a great acrobat or one of those famous car chases in which the hero somehow avoids tragedy at every turn.

It says here that if there had not been a successful, societal reconnection between God and humankind, our little world not have made it to the twenty-first century.

Chapter VI

A BRIEF HISTORY OF HUMANKIND

> *According to the science of biology, people were not 'created.' They have evolved, and they certainly did not evolve to be 'equal.' The idea of equality is inextricably intertwined with the idea of creation.... Americans got the idea of equality from Christianity, which argues that every person has a divinely created soul, and that all souls are equal before God.* Yuval Noah Harari, *Sapiens, A Brief History of Humankind.*

The above quote illustrates two competing theories of human cultural and religious development. In his 2011 book, from which I borrowed the title for this chapter, Yuval Noah Harari discounts the notion that equality of all races, genders and creeds is a necessary condition of our species. He buys the classic evolutionary biologist's argument that all species, including the human species, are the random results of natural selection.

Humans, according to that theory, are exactly like animals: They struggle to survive and procreate, in a "dog-eat-dog" battle in which the fittest survive and the less fit become extinct. In that kind of system, there is no equality whatsoever.

Christianity, and its predecessor religion, Judaism, proclaim a wholly different view of humankind. It is the view that has prevailed in the West from the time of Emperor Constantine to present-day America. (Note that I track the advent of Christianity, as a societal force, from the conversion of Constantine and not from the reported crucifixion or alleged resurrection of Jesus.)

In this chapter, I track the societal and cultural evolution of humanity from a state of inequality, as was found during the decline of the Roman empire, to a professed state of equality before God and before the law. My narrative is not in the form of a convincing argument, but rather a brief description of the various eras, tending to show a growth in tolerance, fraternity and a basic equality of dignity of all citizens, coupled with an equal opportunity to prosper.

At the end of the chapter I conclude that the advent of Judaism, some five thousand years ago, and Christianity, some two thousand years ago, were not only key moments in the history of civilization, but perfectly timed to avoid catastrophe. I will invoke a quote from Einstein, when he said that "God does not play dice with the universe" to support that argument.

It is a difficult argument to win – a difficult thesis to prove. But instinctively, at least, I think my reader will reach the same conclusion by the end of this chapter. That conclusion has already been stated, but it bears restating.

It is as follows.

Had there not been a paradigm change in humankind's destructive tendencies, our species would not have survived into the twenty-first century. Had there not been a steady, if bumpy, ride from the beginning of recorded history through the enlightenment of the Renaissance, followed by the arrival of nearly universal freedom and democracy, our planet would not have survived.

I offer my reader a cursory review of that history. It shows a relentless path from a dog-eat-dog form of inequality and oppression to the globalization of ideals and fraternization of diverse peoples.

Gilgamesh: Beginnings of Recorded History

One of the oldest narratives in recorded history (and about the only that contains a long, coherent story that scholars have been able to decipher) is *The Epic of Gilgamesh*.

We can think of Gilgamesh as the first of the patriarchs – predating Abraham by a few hundred years and the Egyptian pharaohs by a least one millennium and Alexander the Great by about three. We're talking more than three thousand years before the one they called the Christ.

But Gilgamesh was not a lovable patriarch. He had few, if any character traits that we would want our kids to have. He was no saint. In fact, he was the very definition of what today we would call the ultimate sinner: violent, narcissistic, misogynist.

History describes his birthplace as being in the city named Uruk, located in the area of Mesopotamia, that rich region whose pastures were watered by the two famous, diverging rivers: Tigris and Euphrates.

I recently re-read the Gilgamesh epic looking for some trace of nobility, something around which to build culture and civilized conduct. I could find none. The virtues that were extolled were confined to size and strength. It all came down to two fierce men who vied for control of the kingdom. One was Enkidu, who had been raised in the wilds, with animals. The other was Gilgamesh, who was more urbane.

They fought to see who would rule the roost. Here's how the fight is described:

> *[Enkidu] put his foot and prevented Gilgamesh from entering the house, so they grappled, holding each other like bulls locked together. They shattered the doorposts and the walls shook. Gilgamesh bent his knee with his foot planted on the ground and with a turn, Enkidu was thrown. Then immediately his fury died.*

At this point, Enkidu sees that Gilgamesh is the better man, because he is stronger. He bluntly acknowledges it with the following proclamation, in which a god (Enlil) and a maternal goddess

(Ninsun), are given credit for shaping Gilgamesh into the ultimate fighting machine:

> *There is not another like you in the world. Ninsun, who is as strong as a wild ox in the byre, she was the mother who bore you, and now you are raised above all men. And Enlil has given you the kingship, for your strength surpasses the strength of men.*

In all of the Gilgamesh epic, there is hardly any mention of women, except insofar as they may have come to bear one of the men in the story. About the only explicit mention of women, or their role in society is in a passage that describes their role in the taming of Enkidu.

The Women of Ur

Here's all I could find about the status of women in the Ur society; it was a scene in which Gilgamesh instructs a "trapper" to set a trap for his rival Enkidu:

> *Trapper, go back, take with you a harlot, a child of pleasure. At the drinking hole, she will strip, and when he [Enkidu] sees her beckoning, he will embrace her and the game of the wilderness will surely reject him.*

As it happened, the trapper did just that, and Enkidu was seduced away from his wilderness ways. But no further mention is made of the harlot, or of any other female except for his mother, Ninsun, who is a minor goddess of wisdom. (Similar to the Greek muses.)

Before the Hebrews make their appearance in history, the status of women was somewhere between subservient and slaves. In Nordic regions well into the modern era, they were sacrificed on the funeral pyre when their masters died. In India, they met the same fate. In Rome, servant women were just a bit higher than a common slave and basically competed for the sexual appetites of the master, along with the official wife and as many concubines as the particular era allowed.

The Hebrews changed all of that. All of a sudden, there was a sense that a woman could not be discarded by her husband willy-nilly. Single women could be coveted, but once married, they had an esteemed place along their husbands – at least if they behaved properly. They weren't exactly equal to the husbands or fathers, but they weren't slaves or sex-objects.

"The Gifts of the Jews"

The history of the Jewish people, as we all learned in school, started with the era of the patriarchs. Afterwards came the kings and the prophets, followed by the priests, whom today we call "rabbis," a word that actually means "teacher."

Each era has one poignant passage, and we identify the first patriarch with the very touching, if slightly twisted and unbelievable story of Abraham and his firstborn, Isaac.

It took me a while to figure out (on my own) what turns out to be most likely the correct version of the story of Abraham sacrificing Isaac.

For many years, as I thought about it, my logical mind did not accept the idea that (1) an infinite creator would ask a human being to kill and burn his own son to please him; and then, (2) change his instruction and send an angel to countermand it, when satisfied that the human being was so loyal to his creator that he would actually carry out such a cruel mandate.

My legal mind rejected it just as well. For one thing, if you are given direct instructions by the principal, his agent cannot countermand those instructions. Let's say the president of your client's company tells you to do something, but as you are getting ready to follow his instructions, a vice-president calls you and says to do something different. You would stop, of course (just in case). But you would ask for some form of direct communication from the president – or at least an explanation as to why the president changed his mind in midstream.

If the instruction received directly from the president was to buy or sell stocks immediately, you might well insist on hearing from the president directly before changing your course of action.

I would. And I am sure a prophet as wise as Abraham would too.

I was pondering these things about five years ago, when I read a letter to the *Miami Herald* in which a rabbinical scholar argued the same, exact point. What was happening, he said, was that it was a custom in the Middle East to sacrifice children as a way of pleasing the gods. G-d's supposed intervention into Jewish history, represented by its first and greatest patriarch, was meant to stop that practice.

Abraham did not (and could not) receive a command from an infinite being instructing him to kill and burn his son. Abraham was presumably connecting in some unprecedented way with G-d, and felt compelled to do him/her homage, using the best practice of his era. It is precisely an indication that he had some special connection, when we read that he received some sort of divine message not to engage in child sacrifice.

That could not possibly please God.

The lesson of the angel telling Abraham not to kill and burn his son was G-d's way of telling our forebears that human sacrifice was not a good thing. (Or, I should say, G-d's way of telling our male ancestors that – as I suspect that mothers knew that all along.)

Yet it was the custom of the times.

Human Sacrifice in Recorded History

History suggests that in many cases, human sacrifice, dressed up as a religious ritual, was a practice driven by many non-religious factors. For those who were kings or emperors, killing their offspring had all kinds of practical consequences. The main one, of course, was to reduce the temptation that the offspring would kill them and take the throne – perhaps aided by a power-lusting mother or other relative.

More commonly, the practice of human sacrifice was a way of reducing the population when it was felt that some citizens (children in particular) were a burden, or were consuming too much of the available resources. Before the Judeo-Christian tradition took hold, beginning about five thousand years ago, it was a dog-eat-dog society where the strongest ruled and everyone else was expendable.

Women and children were particularly expendable, due to their relative physical weakness. For example, in the oldest civilization for which we have records (the Sumerians in the area lying between modern-day Iran and Iraq), the strongest man – the most alpha of the males – ruled with an iron fist. He had his choice of food, women, and slaves. (Which is a bit repetitive, since women and children were essentially slaves of the men in those days.)

The barbaric idea that we could please the almighty creator by killing innocent children, or burying the living wives of great leaders with their husbands, was part and parcel of what went on in the world when a small tribe of Middle Eastern Semites arrived and put an end, in their small tribe, to that practice.

It took a while for this new sense of ethics to spread to all corners of the world.

Asian tribes that migrated North all the way to Siberia and ultimately crossed ice bridges to the New World preserved the barbaric practice well into the last millennium. Here's a vignette from Yuval Harari's *A Brief History of Humankind*:

> *The Ache people, hunter-gatherers who lived in the jungles of Paraguay until the 1960's offer a glimpse into the darker side of foraging. When a valued band member died, the Ache customarily killed a little girl and buried the two together.*

Hunter-gatherers like the Ache people roamed the Middle East and Asia Minor until perhaps 9,000 years ago, when agriculture allowed humans to live off the land, and built increasingly sturdy roofs over their heads.

Those which stayed in the Mediterranean area learned agriculture, metallurgy and a measure of compassion towards the weakest and smallest members of the species. Human sacrifice was cast into the ash heap of history, as we shall see.

Before the Hebrew Code, Hammurabi's

Approximately 4,000 years later after the hunter-gatherers, the Middle East had seen agriculture combined with domestic animal husbandry to generate fairly stable societies. One of those was the Hebrew people; they forged a very special culture, centered around a perception that they were chosen by single, almighty, personal g-d and protected in their beliefs by doctrinal inspiration and an occasional miracle, such as might be necessary to bring destruction to those who would destroy them.

The doctrinal inspiration carried a rather specific code of conduct. It was a radical departure from everything that came before, including the famous Hammurabi's Code (c. 1790 BC). That code, celebrated as it was, actually enshrined the most odious inequality. Here's Harari's description:

> *Hammurabi's Code asserts that Babylonian social order is rooted in universal and eternal principles of justice, dictated by the gods. The principle of hierarchy is of paramount importance. According to the code, people are divided into two genders, and three classes: superior people, commoners and slaves.*

Our modern sense of equality would never accept such a code. A hierarchy consisting of three classes of people is only quantitatively different from the Roman hierarchy, which consisted of four basic ranks: patricians, tribunes, plebeians and slaves. But it gets worse, when you consider criminal provisions of Hammurabi's Code:

> *The code also establishes a strict hierarchy within families, according to which children are not independent persons, but rather the property of their parents. Hence, if one superior man kills the daughter of another superior man, the killer's daughter* [not the killer] *is executed in punishment.*

Harari, like most mainstream academics of today, is unable to pass judgment on the code; for reasons that boggle the mind – and that stem from a dogged refusal to acknowledge that some cultures and some religions are superior to others, he tries to justify the unjustifiable, saying:

> *To us it may seem strange that the killer remains unharmed whereas his innocent daughter is killed; but to Hammurabi and the Babylonians this seemed perfectly just.*

That's what's called moral relativism; it is rampant, as a theoretical construct, in academia. It is also rampant in the two occupations which I practice: law and politics.

And it's even worse in certain areas of industry, such as banking. As explained by this author in an earlier book (*The Wealth of a Nation*), investment bankers have created for themselves a private game of "Monopoly," with real money, that cheats the ordinary citizen out of a good portion of his savings and rubs in the face of hard-working Americans the super-rich lifestyles obtained by investing "other people's money" with huge leverage factors.

They justify it to themselves (and the larger society) as right because it is legal. Down deep inside they know and we know their wealth is based on immoral practices.

But I have gotten way ahead of myself. The natural law – meaning the universally accepted code of decency – took a while to develop. Significantly, in many instances, the advance of morality had a religious origin.

Such was the case with Moses and the so-called "Ten Commandments."

The Ten Commandments

Humanity received a jolt of ethics from the arrival of Moses and his Ten Commandments. Of all the biblical teachings and laws, of all the stories and characters and moral lessons they convey, there is no single set of rules of conduct that can rival the "Decalogue," or Ten Commandments.

I got my law degree from Harvard University, and have read a lot of cases and treatises in my academic and professional life, spanning close to a half century. In that time, I have found no better code of conduct, no normative recital of civility, no summary of the most important rules for societal wellbeing than the Decalogue.

Historian Tom Cahill agrees with me; and his poetic evaluation of the importance of the Ten Commandments is worth quoting:

> *If I can peer through the mists of history and see the begrimed, straightforward faces staring upwards towards the terrors of Mt. Sinai and if I can imagine the immense throng of simple souls trudging through the whole of history – all the ordinary people down the ages in need of moral guidance in all the incredibly various cultures and situations that the planet has known, it must be admitted that it would be fairly impossible to improve on the Decalogue as we have it. The sins it catalogues are the great sins and those it does not mention explicitly – such as withholding sustenance from those who have nothing – can be deduced from it.*

Christianity took the lessons of the Ten Commandments and added a new commandment: to love all members of our species, even if they were not our neighbors. And there was a corollary to that: It taught that the most exalted form of godliness, the highest degree of virtue, the most perfect expression of the human condition, was achieved when we could forgive those who were doing their best to destroy us.

The spread of the Gospel message to the Roman provinces – far from bringing down the Roman empire – allowed it to bring a measure of civilization to the barbarians in the North and East. Instead of spreading *Pax Romana* by force of conquest, the Christian missionaries spread peace and harmony through their words and example.

British historian Christopher Dawson explains the impact of Christianity on what would otherwise have been a truly barbarian-dominated era, referred to by many as the Dark Ages (c. 400-1200):

Primitive Europe outside the Mediterranean lands preserved no common centre and no unified tradition of spiritual culture. The people of the North possessed no written literature, no cities, no stone architecture.

They were, in short, 'barbarians'; and it was only by Christianity and the elements of a higher culture transmitted to them by the Church that Western Europe acquired unity and form.

It was not a sudden process, of course, to go from barbarians sacrificing their young to Renaissance astronomers (like Galileo) examining stars with their telescopes or monks (Mendel) doing genetic experiments with peas. It took the better part of a millennium, and a rather formidable combination of what others have termed "guns, germs and steel." (More on that later.)

Arguably, the first clear sign that there was a new world order, in which all of our species were considered to have some relevance and worth, happened before the break-up of the Roman empire. It was about four centuries into the birth of Jesus when someone, other than the mightiest soldier with the largest army, was able to infuse benevolence into a temporal ruler.

The episode is worth reviewing in some graphic detail.

A Bishop Stares Down an Emperor

It took place barely four centuries after the birth of Jesus, in the year 390. It was remarkable for all kinds of reasons; mostly it was remarkable because it was the first time that a bishop stared down an emperor.

Throughout recorded history, there were examples of temporal rulers consulting shamans and priests and the prophet-du-jour. Kings and sheiks, chieftains and emperors asked their spiritual advisors all kinds of questions. Some probably followed their advice and even changed their initial, instinctive plan of action.

But in recorded history, no ruler of men had ever been dressed down by a spiritual leader. Certainly not in a public, humiliating way, as St. Ambrose did to Emperor Theodosius.

From the standpoint of political science, the adversarial contest between Emperor Theodosius and Bishop Ambrose is a remarkable event. It depicts a simple bishop interacting with a very powerful politician over the dignity of human life.

And the politician was not just some mayor or governor. It was an emperor – a ruler who had the power of life and death over anyone in his realm, including a lowly bishop. In the incident at hand, the emperor had ordered the summary execution of a few thousand citizens in a region (Thessalonica) whose leaders had defied him.

As cruel as that sounds to us today, it was not unique in the history of the times. Nor was it the worst of such incidents of massive cruelty by emperors. But things had changed, since Constantine had embraced Christianity. The emperor was no longer the absolute, divinely instituted authority.

Ambrose, as bishop, had no temporal power whatsoever. His power to correct the emperor was strictly the power of the word, and whatever spiritual authority could be wielded by a cleric with no army, no weapons, no temporal authority.

Theodosius (Roman Emperor from 378-392) was in many ways an extraordinary emperor. He had successfully dealt with the Goths and other tribes and brought greater unity to the troubled Empire in the West. But he was also famous for having a bad temper.

Here are the details of the massacre, followed by the abject, public repentance of the emperor.

Massacre in Macedonia

It happened in Thessalonica, which is in modern-day Greece (though it was then known as Macedonia). A riot had broken out involving the Gothic captain of the Roman Garrison (Botheric) and a very popular charioteer, who was also a drunkard.

Botheric had the charioteer arrested for debauchery. The crowds in turn rose up in favor of the athlete and rioted. The melee had ethnic implications, as the town was largely Greek and the garrison captain was a Goth.

In the melee, Botheric was killed.

The emperor went berserk and gave an order to convene the citizens into a stadium and then massacre them all. After sending the order, he tried to countermand it, but the message canceling the order did not arrive in time. In the end, about 7,000 people died in the carnage.

Theodosius was mortified and went to Milan to seek solace from St. Ambrose. But Ambrose, fearing the Church was just being used as a political tool, refused to meet him and instead wrote a lengthy letter chastising the emperor.

The letter contains three key elements: declaration of respect for the civil authority, chastisement for the ruler's immoral conduct and a call to public repentance.

The letter also cites past incidents in which Hebrew rulers – most notably King David – had been forced to repent from cruel and mischievous actions. An excerpt that illustrates the first element went as follows:

> *Listen, august Emperor. I cannot deny that you have a zeal for the faith; I do confess that you have the fear of God. But you have a natural vehemence, which….if any one stirs it up, you rouse it so much more that you can scarcely restrain it……would that….no one may inflame it!…. restrain yourself, and overcome your natural vehemence by the love of piety….*

The second element (condemnation) is illustrated by the following excerpt:

> *There was that done in the city of the Thessalonians of which no similar record exists, which I was not able to prevent happening; which, indeed, I had before said would be most atrocious when I so often petitioned against it, and that which you yourself show by revoking it too late you consider to be grave, this I could not extenuate when done. When it was first heard of….there was not one who did not lament it, not one who thought lightly of it; your being in fellowship with Ambrose was no excuse for your deed…..*

That is followed by a plea which incorporates the analogy to King David – thus conflating the legacy of the Old Testament with the emergence of Christian influence in civic affairs:

> *Are you ashamed, O Emperor, to do that which the royal prophet David, the forefather of Christ, according to the flesh, did?....he said: I have sinned against the Lord. Bear it, then, without impatience, O Emperor, if it be said to you: You have done that which was spoken of.... say: I have sinned against the Lord. If you repeat those words of the royal prophet: O come let us worship and fall down before Him, and mourn before the Lord our God, Who made us. [I]t shall be said to you also: Since you repent, the Lord puts away your sin, and you shall not die.*

Ambrose ends his missive with a lyrical appeal, also based on the Hebrew bible.

The quote from Ecclesiastes is one that I have used in many occasions. It is a beautiful, literary gem that seems to transcend every age and fit every circumstance:

> *To everything there is a time, Ecclesiastes 3:1 as it is written: It is time for You, Lord, to work. It is an acceptable time, O Lord.... For the Word of God Himself tells us that He prefers the performance of His commandments to the offering of sacrifice. God proclaims this, Moses declares it to the people, Paul preaches it to the Gentiles....*

Emperor Theodosius accepted the rebuke and did public repentance. History tells us that he even wore sackcloth and ashes, abiding by the custom of the times.

It was a watershed in church-state relations. From this point forward, and for the next thousand years, emperors and bishops (particularly the bishop of Rome, also known as the pope) jousted and wrestled and occasionally even fought in the fields of war.

Popes and bishops didn't always prevail in the confrontations. But there was a general sense that kings and emperors were not gods and did not have absolute power.

It was a clear, distinct step towards individual freedom and equality of all citizens before the law. The idea that each and every human being has infinite value as a creature of God, that rulers are not superior in spiritual worth, but only in temporal authority, and that no single member of our species, no matter how small or weak, is expendable, began to sink into the collective consciousness of Southern Europe, the Middle East and North Africa.

Modern scientists have focused on that region of the world as the one that contained the best components of what was needed for progress and prosperity. One of those was Jared Diamond. He called it the "fertile crescent."

Guns, Germs and Steel

At this point, I need to digress and delve a little bit into a sociological model that is in vogue; its basic tenet is that the advance of civilization is centered in the so-called "fertile crescent," comprised of the corner of the Mediterranean that connects Eastern Europe with Asia Minor and the Middle East. That area is, by all accounts, the earliest part of the world to see agriculture, the use of basic metals (copper and iron) and organized clusters of shelters and markets, using collective utilities such as aqueducts.

The argument in favor of the fertile crescent model was formulated rather convincingly by Jared Diamond in the celebrated book *Guns, Germs and Steel*, which is now a sort of an anthropological bible.

And it's mostly true, but missing an important element. The fertile crescent was the cradle of civilization for the mentioned reasons, but also, and perhaps most importantly, because it was the venue for a new, important factor: faith.

If you go back to the first stirrings of human consciousness, you will find, in every case, three things. One is the discovery of fire; after that is the use of a very sharp, naturally occurring flint stone produced by volcanoes and called "obsidian."

The third and equally important element is faith.

Fire, flint and faith. That is the stuff of civilization.

In summary, we have the discovery of fire, followed by the use of sharp obsidian flint-stones to make hides that could cover the mostly bare human body, and the faith-based recognition that all humans, all races, all ethnicities, are equal in dignity and worth powered civilization from that moment forward.

I would replace Diamond's "guns, germs and steel" with my own triad: "fire, flint and faith." For it would be fire, flint and faith that would mold Western civilization, spreading North and East during the next thousand years.

What the Roman empire could not bring about by the use of physical force was brought about by a spiritual force that seemed to mix in with an indigenous, innocent, ubiquitous inclination of even the most barbarous tribes of the North (Lombards, Saxons, Franks, Celts, Suevi) and East (Huns, Goths, Magyars) to look to heaven for moral guidance.

But that came later. First came Constantine. For my readers who don't know the history of the early Christians, one vignette I can share is found under St. Peter's basilica. Here is the tomb where the first popes were buried. Next to each name, for about the first two centuries A.D., are the letters "s" and "m."

Let me tell you what the letters stand for, in case you don't know: "saint" and "martyr." Yes, you read right: For a couple of centuries after Jesus walked the earth, his followers were martyred just for believing and preaching the Gospels.

But that wasn't all. The heads of those who professed the Christian faith were placed on top of stakes, doused with a flammable material, and used as lamps to light the roads. And to dissuade others from following their example.

Emperor Constantine ended, in one fell swoop, the persecution of Christians. The story of how an emperor became a believer, by motherly persuasion, is there for all to read. This book is not about faith. It is about facts.

And the fact is that Constantine changed the course of history.

The Rise and Fall of the Roman Empire

Before Constantine, the Christian era was floundering. And so was the Roman empire, which by that time had conquered faraway lands.

Science and the Theory of God

Gaul and Hispania and Britannia were frontier regions; and they were reasonably pacified and Romanized.

It was called *Pax Romana*.

But *Pax Romana* was about to be fragmented, by barbarians from the North and East. What saved Europe is that the barbarians had been slowly converted and had begun to appreciate the wonders of Judeo-Christian civilization.

Europe, during and after the fall of Rome, became the cradle of civilization, which was later the springboard for freedom and democracy in the whole world. Freedom and democracy would never have become universal forces had Christianity not overtaken the native, barbaric, intolerant culture of the North and East.

Instead, we might have had a continuation of what has been called the "Dark Ages." From the fall of Rome in the fifth century to the beginning of the Middle Ages, almost a millennium later, the advance of civilization was slowed down by invasions from the North and East of Europe. Much better historians than I (including Jon Meacham and Thomas Cahill) have disputed the notion that these were "Dark Ages" at all.

But it was certainly tough sledding.

No one denies that Rome was sacked and conquered repeatedly and that the empire broke up into an East and a West, with the West basically a battleground of barbarian invasions, followed by Muslim expansion. The Magyars, in modern-day Hungary, were not so much conquered as converted; that happened in 1,000 A.D., when Stephan I became the first Christian king of Hungary.

Muslim invaders were not completely repelled until halfway through the second millennium, when they were finally beaten decisively at Lepanto (1571) and Vienna (1683).

But the prior millennium (c. 500 A.D. to 1,500 A.D.) was not a wasteland of wars, plagued by ignorance and disease, as some writers have portrayed it. To the contrary, what took place in Europe, the Middle East and Northern Africa was an inexorable expansion of Jewish and Christian scholarship, preservation of Arabic culture and learning (particularly in the areas of math and science), and amalgamation of native cultures into a common, diverse but fundamentally unified civilization.

By accepting Christianity, Constantine had rejected idolatry and all the cruelty that accompanied it, including animal sacrifices. The notion of human equality, that all men and women are children of a benign creator, began spreading until it was ultimately accepted everywhere in the West.

Intellectually, the interdependence of faith and reason was set early on by Augustine, Bishop of Hippo in Northern Africa (modern day Algeria) when he said: "Believe, so you may understand; understand so you may believe."

Good stuff, and intellectually sound, but intellect was not what changed the barbarians. They were changed by the example of monks and friars, followers of St. Benedict, St. Patrick and Jesus, who took it upon themselves to bring not only Christianity, but civilization to barbaric lands of Northern and Eastern Europe.

How the Monks Saved Civilization

One can only marvel at the constancy, fearlessness and altruism of the monks, friars, nuns and brothers – all religious celibates – who were known individually as "clerics" and collectively as "clergy." During the thousand years after Constantine, these folks were a combination of social workers, teachers, horticulturists and researchers whose work and conduct exemplified Jesus's startling pronouncement that required societal leaders to "serve" rather than "be served."

Service, and the preaching of the Gospels – a message strictly of love and forgiveness – often merited death for the messenger. And the required response from these peaceful missionaries was to "turn the other cheek."

How strange, how totally unprecedented; how transformational.

Christianity accepted the Jewish Decalogue and added a new commandment: to forgive those who have wronged us. Surprisingly, the message of meekness and forgiveness ultimately took root, even among barbarians of the North, East, and South.

Here's a vivid description of the impact that Christian missionaries had on the barbarians of the North of Europe, from the pen of British historian, Christopher Dawson:

> *The lives of the saints and ascetics impressed the mind of the barbarian because they were the manifestation of a way of life and a scale of values entirely opposed to all they had hitherto known and accepted.*

Dawson describes in vivid detail a magnificently humble infusion, by pure example, of the work ethic, combined with respect for nature, a zest for scholarship and a unique, novel spirit of sharing:

> *Silent men were observed about the country, or discovered in the forest, digging, clearing and building; and other silent men, not seen, were sitting in the cold cloister, tiring their eyes and keeping their attention on the stretch, while they painfully copied and recopied the manuscripts which they had saved.*

The impact of radical Christianity, as preached and exemplified by these peaceful, "silent" men was quite startling. Says Dawson:

> *By its sanctification of work and poverty it revolutionized both the order of social values which had dominated both the slave-owning society of the Empire and that which was expressed in the aristocratic warrior-ethos of the barbarian conquerors....*

The result was a whole new perception of equality such that "the peasant, who had for so long been the forgotten bearer of the whole social structure, found his way of life recognized and honoured by the highest spiritual authority of the age."

Needless to say, the transformation of society that culminated in the French Revolution slogan of "*liberte, fraternite, egalite*" took another few centuries to materialize.

The Middle Ages

During the Middle Ages (roughly 1,200 A.D. to 1,600 A.D.) Europe was transformed from a feudal society, in which there were two classes of people (the nobles and the serfs) into an increasingly egalitarian society.

Intellectually, science and religion were companion tools in the human search for truth. Occasionally, there were conflicts, such as the celebrated one that pitted those who believed the earth orbited around the sun versus those who believed the earth was the center of the universe.

The protagonist of that conflict was Galileo.

Galileo's battle with the Church of Rome has been featured in so many popular books and movies that most people probably think he was tortured for his views, when in reality he lived the life of a nobleman, and poked fun at the very authorities that disagreed with him. (The book to read is *Galileo's Daughter,* by Dava Sobel.)

Nor did anyone really think the world was flat. Greek and Egyptian astronomers had been clear in their understanding of the general relationship of the moon, the sun and the stars. It remained only for the Polish polymath, Nicolaus Copernicus (c. 1532), to quantify orbits and describe them with rather startling precision that astounds us even today.

It took a while for the predominantly Christian populace to accept the idea that the earth was not the center of the universe and that the sun did not revolve around the earth, as the Bible seemed to imply. They should have listened to Augustine when he said, early in the prior millennium, that if the Bible seemed to contradict science, we were probably interpreting it wrong.

Ultimately, the skeptics lost the evidentiary battle. Science was allowed to proceed with its investigations; and religion – at least within the broad outlines of Christianity – was also given broad scope.

It didn't happen overnight. There were excesses, and the world learned an important lesson: it is not a good idea to combine religious power with temporal power. In the meantime, in the civic arena, the masses were clamoring for some measure of equality. England was the unsurpassed leader, with the landmark proclamation of the Magna Carta.

Science and the Theory of God

The Magna Carta (1215)

It took eight centuries, from Constantine to King John of England, for a temporal ruler to admit that he had to share power with the people as a whole. After that, the momentum changed, and the Western world slowly evolved towards equality.

It was in 1215 and the document that was drafted and presented to the king still inspires lawyers and non-lawyers everywhere in the world. It was called the Magna Carta.

Three centuries later, in 1598, one of the most enlightened countries in Europe proclaimed that humans had an inherent right to choose their belief system. This was the last link in the chain that brought not just material and political equality, but also religious equality to the Western world.

It was called the Edict of Nantes, and it happened almost half a millennium ago. It should be noted that the idea of religious freedom still has not been accepted in many regions of the Islamic world; and hence, the tension we feel in a number of Islamic theocracies of the Middle East and North Africa.

Other than Islamic theocracies, the only remnants of intolerant regimes are those which still subscribe to atheistic Marxism. Of those, China is the only serious world player. It is approaching the United States and Europe in industrialization, but it is not even close to being the world's cultural leader, let alone the leader in terms of guaranteed levels of prosperity and liberty.

Leading the world in terms of enlightened, progressive conditions is a coalition of nations that have accepted the Judeo-Christian prescription and have added a much broader form of Greek democracy. How Europe and the United States came to be the catalysts for a worldwide coalition that has promoted human dignity and participatory democracy, often at the cost of lives and treasure, is a story in itself.

We will get to it soon, but first a brief story of religious tolerance.

The Seeds of Religious Tolerance

Once Christianity became Christendom, and its spiritual power was mixed with temporal power, the traditional problem occurred. They say, "power corrupts and absolute power corrupts absolutely."

In the Middle Ages and particularly in Spain and Italy, the notion that spiritual truth must guide all aspects of public policy took hold. Contrary to some popular accounts, it was not persecution of diverse faiths *per se*. For the most part, the Jews, Protestants and Muslims in Europe were not bothered in their practice of a separate religion, or made to swear allegiance to Christianity or Catholicism.

That is to say, they could pray and practice as they wished in the privacy of their homes. And, in general, they could dress as they wished in the public square. In the main, their temples of worship were respected.

Preaching in the public square – and particularly any inflammatory speech that chastised the clerical authorities – was not treated lightly. But it was nothing like the imposition of Sharia law in Islamic states, the torture of Jews in Babylonian times and the slavery imposed by Egypt on the Hebrew nation. It certainly did not involve the cruelty directed at followers of Jesus during the first three centuries of Christianity.

It was more like the modern maxim of "don't ask don't tell."

Yet, all the time, inexorably and progressively, despite occasional spasms of anti-Jewish sentiment or Inquisition-style persecution of non-believers, Europe moved incrementally towards tolerance and equality before the law.

Contrary to what some writers (and the popular media) have conveyed, the Christian nations were not the *least* tolerant, but the *most* tolerant. While it is true that Jews were expelled from Spain in 1492, within one century the more enlightened Spanish clergy was embracing them as equal to Christians.

One priest, who founded the *Escolapio* (meaning "pious schools") teaching order and who was later canonized, is worth mentioning in this context. His name was St. Jose de Calasanz (1557-1648), and he started the first truly public school in Europe. That is remarkable in itself, for in that era schools were still reserved for royalty and the wealthy.

But Jose de Calasanz went further in his unique embrace of diversity: He insisted that Jewish children be taught for free, alongside Christian children. (I am proud that his mother had the same last name as my mother, Gaston, and that he was my lineal ancestor....)

Ultimately, and propelled by the very doctrinal force of the Gospels, religious tolerance was proclaimed throughout Europe – to the point that the entire region is currently suffering the onslaught of immigrants who, in many cases, seem intent to destroy the very culture that welcomes religious differences.

I should add that historians generally point to an American colony as the first to establish absolute religious freedom. It was in what we now call the State of Rhode Island and was proclaimed by Roger Williams (1636).

Once again, America was an exceptional nation.

The Last Four Hundred Years: From Renaissance to Modernity

After the Edict of Nantes and the expulsion of foreign invaders, it is fair to say that Europe flourished. Much has been written about the era; it is known as the Renaissance, when arts, letters and astronomy flourished like never before.

Democratic movements also flourished; and so did economic arrangements between labor and owners. The most notable was in agriculture, which at the time involved close to 100% of the workers. The only citizens not employed in the agricultural sector were the rulers, who were almost always of royal lineage, and the clergy; tradesmen, shop-keepers, and blacksmiths, plus a handful of artists and academics comprised a small minority, which later became identified as an important sector of society, called the "bourgeoisie."

The discerning reader should keep in mind the above categories, which acquired a special semantic importance when they were collectively referred to as the three "estates" of society and incorporated into the French ruling body called the "Estates General."

Knowing this is very helpful to the American reader, who might have been puzzled when hearing the media call itself "the Fourth Estate." How it is that journalists separated themselves from the other three sectors of society is a story in itself. (The book to read is by this author, *Democracy in America 2010*.)

The last important era of the last half-millennium was characterized by a remarkable, unprecedented, exponential growth in technology and industry. Europe, followed closely and in some cases led by North

America, experienced a fantastic growth in industrial production. Electricity, plus the steam engine. followed by the internal combustion engine, the cotton gin, and every kind of harvesting machine and metallurgical innovation combined to transform the face of Europe.

It was called the "Industrial Revolution." The machines invented in that period – notably the steamship – enabled Holland, Spain, Germany, Portugal, and England to establish colonies in Asia (particularly India and China), Africa and all of the Americas. The colonies were provided with the same industrial machines that were invented in Europe and North America.

Adding to the transformational power of the Industrial Revolution was an explosion of medical and pharmaceutical technology. This explosion of technology extended human longevity and quality of life, which in turn had a multiplier effect on the two factors that spur economic growth: the number of workers and the productivity of each worker.

Despite some clear disparities in wealth and income, the proverbial "rising tide" lifted all ships, big and small. Equality of economic condition began to accompany philosophical and theological pronouncements that equated all humans as to dignity.

By 1776, Americans were affirming that "all men are created equal." Soon after that, in 1789, French revolutionaries were proclaiming not just *liberty* and *equality*, but even *fraternity* of all citizens.

The stage was set for the coming of "one-man-one-vote" democracy and, ultimately, one-person-one-vote.

But first, two malignant tumors had to be extirpated from the body politic. One festered mostly in the first half of the twentieth century; the other in the second half.

Before we deal with the two macabre elements of what should have been a happy century (the twentieth), let's review the spasmodic, and yet relentless rise of equality that has been the emblem of the last 250 years in the history of humankind.

The Rise of Civic Egalitarianism

The last two-and-a-half centuries were like the icing on the cake that completed a process by which the bulk of the world's population embraced individual liberties and the right of self-determination; the latter feature usually led to a form of democracy, in which those who govern and those who are governed are consulted equally before a legal obligation can be imposed. (That, parenthetically, is the definition of democracy; the document to read is the 1942 Christmas Message by Pope Pius XII, called *Benignitas et Humanitas*.)

By the end of the eighteenth century, there began to emerge, particularly in the West, a consensus on the fundamental rights of every citizen. France and the United States led the way, though France quickly convulsed into anarchy, followed by dictatorial and imperialistic empire.

In the United States, there were two revolutions in less than eight decades: One led to independence and democracy and the second led to equality of all races. *By 1865, America was the worldwide symbol of equality.*

Another three quarters of a century passed before the American model was expanding to every corner of the earth. A hundred nations convened in 1948 and pronounced a universal declaration of human rights; presiding over their deliberations was none other than the wife of the U.S. president. Her name was Eleanor Roosevelt and her two most fervent disciples hailed from two Caribbean islands.

I owe Harvard Law professor and former U.S. Ambassador to the Vatican, Mary Ann Glendon, an understanding of the extraordinary alignment of stars that led to what is, without question, the most significant anthem of liberty in history. It was inspired to a great extent by Winston Churchill, orchestrated by Eleanor Roosevelt and drafted, in great part, by the representatives of two small islands: Cuba and Haiti. (The article to read, by Ambassador Glendon, was published in the Harvard Human Rights Journal, Volume 16, and is entitled *Forgotten Crucible: The Latin American Influence on the United Nations Human Rights Idea*.)

Besides the fact that the two eminent diplomats (Cuba's Guy Perez-Cisneros and Haiti's Emile St. Lot) represented two small, Caribbean nations, the other significant fact is that they drew intellectual support from the social teachings of the Catholic Church.

I have done some research on the question of how much the initial draft of the U.N. Human Rights Declaration was influenced by the teachings of the Church. The clearest correspondence in texts was in relation to the right to education. Both the U.N. Declaration and the papal teachings (more specifically, the famous encyclical by Pope John XXIII) refer to the "prior right of parents to the education of their children."

"Prior" in this context means "primordial." It means that the state is in a secondary position in delivering educational services to the populace. It means, ideally, that children attend private schools of their choice, or charter schools funded by the state through vouchers, or public schools in which parents' (and teachers') associations have an important say as to curriculum, discipline and values taught.

Those are all logical elaborations of the principle in question. How society taxes its citizens and how it allocates the funds to provide basic education to all children, including the poor, is up to civil authorities – without interference from any church.

But it must be recognized that much of the inspiration that prompted civic leaders to proclaim the right of all citizens to a basic education, as part of their right to a "decent standard of living," came from biblical teachings.

And, I should add, the concept of "social justice" arose as much from Judaism as from its sister religion, Christianity. It is by no means a Christian or Catholic invention, as we shall see.

Jewish and Protestant Roots of Social Justice

Concepts of social justice begin to prop up in the writings of the Hebrew prophets – most notably Amos. His preoccupation with the poor and oppressed is seen in the following passages, in which he chastises Israel for its lack of social justice:

1. The poor and needy are oppressed, denied justice in the courts (Amos 5:7, 10, 12, 15) and forced into slavery (Amos 2:6–7; 8:6).
2. The goods of the poor are confiscated (Amos 5:11), and their garments taken in pledge, in direct violation of Exodus 22:25–7 (Amos 2:8).

3. Trade is dishonest, with prices inflated and crooked weights and measures used (Amos 8:5).

It makes me wish that Amos were around today to help me promote the goals of the "Economic Prosperity Committee," which I chair and which is charged with overseeing efforts to provide affordable housing, health care and economic development in poor areas of my county.

Jewish inspiration was followed by a strong Gospel message of helping the poor. And it wasn't just Catholics; the theological basis of social justice, particularly in the United States, is found in a movement that was in great part Episcopalian. It flowered in the second quarter of the twentieth century, and its chief protagonist was a woman.

Her name was Frances Perkins; she was the first woman to be appointed to the cabinet of the United States of America. Her influence shaped the single most holistic pronouncement by an American president in history.

Ross Douthat and Reihan Salam, in a recent (2008) work named *Grand New Party*, describe the eloquent case that Labor Secretary Perkins brought to bear on the struggle for social justice:

> *The poor people have a right to their homes, the same as the rich, and we should not be allowed to enslave them in a form of industry which refuses not only their liberty, but the wage they ought to have in return for the labor they perform.*

The language was almost literally taken from a startling pronouncement made more than a half century before by Pope Leo XIII (1891); in an encyclical called *"Rerum Novarum"* ("On the Condition of the Working Classes"), Leo XIII laid out a whole framework of fundamental labor rights.

For a couple of years, I had researched the doctrinal trail that seemed to lead (semantically, at least) from the nineteen-century pronouncements of a pope to the New Deal labor secretary. All I could find was the previously mentioned involvement of Guy Perez-Cisneros and Emile St. Lot, both of whom were identified by Ambassador Glendon as being inspired by the social encyclicals.

Then I read Jonathan Alter's *Defining Moment*, written just a few years ago (2007). In that marvelous biography of Franklin Roosevelt, Alter describes a period in the early career of FDR in which he was really an intellectual light-weight. His aide and later State of New York's Secretary of State, named Edward Walsh, had a suggestion for the young politician: read Leo XIII's encyclical on labor rights.

The rest was history, and it was enshrined in a historic proclamation of four basic rights, later depicted by an iconic American painter: Norman Rockwell. (One, depicting "Freedom of Worship," adorns my office at Miami-Dade County Hall).

The transcendental proclamation was called the "Four Freedoms," and it was propelled by the pronouncements of the two strains of Christianity that pervaded the three most liberal nations on the face of the earth: France, Great Britain and the United States.

A little additional research reveals that two others who influenced the drafting of the UN Charter, relying on Christian Democratic principles, were France's Rene Cassin and Lebanon's Charles Malik, with some input by P.C. Chang of China (Taiwan). (The book to read is by Allan Carlson, called *Globalizing Family Values.)*

The Four Freedoms

Whatever their philosophical roots, it is unquestioned that Western European ideas gave birth to the nation that Abraham Lincoln called the world's "last, best hope" and which we know as the United States of America. It was there that democracy took root – as explained by the great, Alexis De Tocqueville. (The book to read is *Democracy in America*.). It was in America that our species witnessed the magic of democracy as a political system that most satisfies the human need for self-determination.

Democracy is clearly important, as it fulfills the human aspiration to self-governance. But the Judeo-Christian message goes much farther than the mere right of self-determination or self-governance. It suggests that in order for a society to be a "just" one, it must also foster the "common good." And the common good requires more than just freedom of opportunity, more than just the right to "life, liberty and the pursuit of happiness."

It requires actual fulfillment of that pursuit. More than just allowing individuals to strive for happiness, the idea is to *collectively guarantee at least a minimum level of welfare for all.* That idea took a while to germinate in the American consciousness. It had been proclaimed in the prior century, but it did not take root in the United States until the middle of the twentieth century.

And it took the bigger-than-life appearance of Franklin Delano Roosevelt to give it rhetorical form.

One of America's greatest presidents, FDR must have sensed that it was time to proclaim not just the negative rights ("freedom from") but at least one generic, positive right. His Four Freedoms did precisely that; and it could not have come at a better time.

By the time he made this transcendent pronouncement, the word was convulsed in the second of two horrific world wars. For the second time in half a century, the United States was forced to come to the aid of Europe or risk having the world's most enlightened continent (Europe) under the boot of an odious, racist, materialistic ideology.

Nazism was beaten back and consigned to the "ash heap of history."

Unfortunately, this was not the last gasp of evil in the free world. There was yet standing one more malevolent movement that could easily have choked any semblance of liberty in our planet.

The twentieth century, which by all rights should have been the most enlightened of all periods, with Judeo-Christian civilization firmly implanted in the most advanced parts of the Old World and in most of the New World, was still struggling to get through a veritable "mini-Dark Age," lasting three-quarters of a century, from 1914 to 1989.

It is worth our time to consider, at least briefly, what philosophical deviations fueled these two movements. The second one (communism) has been analyzed and studied to death in every campus of every college in America.

But the first, and perhaps more malevolent one (Nazism) is worth analyzing in a little more depth. Its theoretical tenets were so bizarre, so hateful, so barbaric, that it is really a wonder that an advanced people, led by a mustachioed demagogue, could come close to destroying all that was good in Europe.

The Philosophical Roots of Nazism

I was born in the mid-twentieth century, which happens to be a period of time that was troubling to both science and to humanity. I was born just about halfway into a period in which the world barely survived two movements that were very hostile to the Judeo-Christian belief system. Both were rooted in materialism; both denied the inalienable rights of individuals; both denied God.

The first (Nazism) preached that a Teutonic race was superior to other races, and had the right to eliminate anyone found inferior, so that the master race could preserve our species. It scoffed at the idea that all humans are sons and daughters of a creator – hence brothers and sisters of one another who should strive to love and forgive.

The second movement (communism) was just as devastating. It preached that humans are not individually valuable, but must abide by the only collective plan that allows all to have a measure of equality. It scoffed at the idea of individual rights, and at its corollary: that one cannot justify those who happen to disagree with the collective plan of action necessary to bring about equality.

Neither Nazism nor communism could have been predicted by what came before. Both contradict the Judeo-Christian idea of the individual person as inviolable and possessed of fundamental dignity. Both elevate the state and its goals over the dignity of the individual and the family.

But communism is rooted in a noble aspiration of people, which is the principle of equality itself. Its theory is palatable enough to Christian writers that it comprises a large portion of what is perhaps the most quoted book to propose an earthly paradise. I refer to Thomas More's *Utopia*, whose title has given us an adjective (*utopian*) that is in common usage.

Nazism has no noble elements. What it does have is a strong component of a philosophy that is ever in vogue, and that bears the title of "Social Darwinism." As we shall see, it is the logical result of science trying to define humankind as a slightly smarter animal, whose only purpose in life is to survive.

It is not pure coincidence that the rise of Nazism occurred just when science was claiming to explain truth without any help from theology, arts or any of the humanities.

Nothing whimsical, or lyrical or sentimental was allowed into the ideological scheme presented by Nietzsche and put into practice by Hitler.

The factors that led to this malignant phase in European civilization have been analyzed by many philosophers, many historians, many thinkers much wiser than me. I defer, therefore, to Will Durant, who with his wife wrote the multi-volume *Story of Philosophy.*

He explained it thus:

> *Darwin unconsciously completed the work of the Encyclopedists; they had removed the theological basis of modern morals, but they had left the morality itself inviolate, hanging miraculously in the air; a little breath of biology was all that was needed to clear away this remnant of imposture. Men who could think clearly soon perceived what the profoundest minds of every age had known: that in this battle we call life, what we need is not goodness but strength, not humility but pride, not altruism bur resolute intelligence; that equality and democracy are against the grain of selection and survival; that not masses but geniuses are the goal of evolution; that not 'justice' but power is the arbiter of all differences and all destinies. So it seemed to Friedrich Nietzsche.*

The philosophy of Nietzsche, superimposed on Social Darwinism, held a certain fascination for Hitler, and it even had an etymological connection to his first name. Here's how Metaxas describes it:

> *The name Adolf is a contraction of the Old German 'Adelwolf," meaning 'noble wolf.' Hitler was aware of this*

> *etymology, and in his mystical and eerie way, he adopted the Teutonic and totemic symbol of the wolf as his own. The wild, carnivorous and Darwinian ruthlessness of the beast appealed to him.*

I wish the world, and particularly Hitler, had never heard of Nietzsche.

I wish, instead, that Germany, after losing a world war, had fallen back on its Judeo-Christian roots, as they did after losing a second world war, under the leadership of Konrad Adenauer.

The Western Paradigm: Combining Two Forms of Justice

Nazism was eradicated by force. Its precepts, including the idea of superior races and of eugenics (the elimination of the mentally or physically handicapped) have been banned from civilized discourse.

Communism has been effectively neutralized by a combination of Western acceptance of social justice precepts and the collapse of the Soviet Union, which just about convinced everyone that free enterprise is a better system of providing economic welfare to the masses than centralized planning and control.

In the final analysis, we can see that the Western European paradigm, blending faith (essentially the Judeo-Christian religions) and reason, has become the unifying belief system of the great majority of the world's inhabitants. In societal terms, its basic tenets are the inviolability of the human person, endowed with certain fundamental rights, and a rule of law that binds all citizens equally but can only be practically enforced if there is a common belief in its intrinsic, moral imperative.

This is an important concept. In order for society to function, the great majority of citizens need to have internalized common values. It is impossible to enforce civilized behavior if a substantial portion of society disagrees with a common code of conduct, such as prescribed by the Ten Commandments, coupled with the idea of social justice.

And it should be noted that our modern understanding of social justice is no more than a reconfiguration of the principle defined by Aristotle and Aquinas as "distributive justice." But distributive or social justice must be balanced by "commutative justice," which is the right of private parties to negotiate a contract to buy and sell goods, services or real property.

Countries that try to guarantee distributive justice without guaranteeing commutative justice end up in some form or totalitarianism. Such is the case of China, which seeks to have a measure of economic freedoms, while eviscerating individual freedoms under a banner of a socialist collective.

The China Model

Despite its apparent economic success, China continues to be an epochal mess. For the better part of the last two thousand years, China has chosen to be isolated, monarchical, and flat-out dictatorial – not to mention backwards. Only by opening up to the West (and taking advantage of both foreigners and its own oppressed population) has China become industrialized.

Yet even now, its economy is unequal, unstable and unpredictable. Its people are unequal, oppressed in the main, and potentially explosive.

Modern, economically capitalistic but politically communist China needs at least one revolution or major reform to reach equality. As recently as 1989, its totalitarian rulers stamped out a human-rights spring – exactly two hundred years after the French (1789) declared a Republic buttressed by liberty and equality.

One wonders whether the West has played with fate in engaging in "ping-pong diplomacy" and free trade agreements with this markedly "evil empire" whose dynastic, overwhelmingly male rulers control even the number of children a family can have. The quandary will be answered in due time. It is such a worrisome prospect that it almost leads me to shed all scientific mantra and simply pray for humanity, if this giant police-state is not converted to some brand of humanism.

My instinct, based on what has so far transpired, tells me that China, too, will be civilized; that its "iron curtain" will fall and its people will embrace the tenets of universal brotherhood and sisterhood. If so, it will be no more than a repetition of what has transpired over the last twenty centuries, since the birth of Jesus.

It is so remarkable, so unexpected, so absurdly unlikely, under the circumstances, that it begs for some other-worldly explanation. It required more than a common code of conduct (the Ten Commandments); it required the kind of collective faith that permits disparate peoples, of different ancestries and languages, to forgive their own historical enmities and unite to fight evil, often knowing that the odds are stacked against them and that the instinct for self-survival dictates appeasement, accompanied by a "separate peace."

Churchill and Roosevelt, Reagan and Walesa and John Paul II did not embrace the strategy of appeasement, coupled with the proverbial *separate peace*. They opted for resistance, powered by an abiding hope in a force beyond human calculation.

And it says here that they succeeded, precisely because of their faith.

Could the World Have Survived Without the Judeo-Christian Paradigm?

Nazism, as an intellectual movement, was presumably eradicated by the defeat of Hitler. Communism was at least partially eradicated by the fall of the Soviet Union and the wall separating East Germany from West Germany.

But not all anti-Judeo-Christian ideologies have become defunct.

Even now, the Middle East is convulsed with war, persecution and lack of religious freedom. In the last decade of the twentieth century and the first decade of the twenty-first, the carnage has been particularly evident in areas where two antithetical strains of Islam inspire the various peoples. One sect is called Sunni and the other is called Shiite. They are by no means the only sects that divide Islam; they are simply the ones that prevail in the majority of Islamic states.

Inside those sects are fanatics who see their sacred mission as eliminating the largely Jewish (but actually fairly diverse) state of Israel and either subjugating or at least removing from their territories any large or influential presence of Western nations like the United States.

Besides the small minority that dreams of installing a new Caliphate, where only Muslims reside and only Islam is practiced, there are occasional wars between the sects (the previously mentioned Sunnis

Science and the Theory of God

and Shiites). Justified by religious fervor, one suspects there is often a more secular agenda, such as the most precious regional resource: oil.

Such was the origin of the invasion of Kuwait, in 1991, by Iraq. Whatever his other shortcomings, George Herbert Walker Bush was up to the task of defending this small nation, which had done no harm to any of its neighbors and which was a beacon of tolerance in an area beset by intolerance.

Crafting together a coalition of nations that had never been united, except in name (under the U.N. banner), Bush the elder surgically excised the Iraqis from Kuwait. It was a feat not equaled since the Bill Clinton-led military intervention of Bosnia, which, ironically, saved the Muslim minority from genocide.

Jon Meacham narrates the stakes for the world if the seizure of a small country (Kuwait), by a neighbor (Iraq) of almost identical culture and religion had been allowed to go unmolested.

Here's how Meacham describes the stakes:

> *By turning the small Persian Gulf emirate into what he called the '19th province, an eternal part of Iraq,' Saddam hoped to transform himself and his nation into more significant players in the Arab world and beyond… It was an untenable prospect for a global economy so dependent on petroleum. The possibility, moreover, of such a wealthy, powerful Saddam was inherently destabilizing given his proven capacities to use weapons of mass destruction…*

Bush obtained near-unanimous support from the U.N. National Security Council (with two abstentions) for the use of an international coalition to liberate Kuwait. He did not follow that with a war of regime change. The subsequent decision by his son, George W. Bush, to carry out regime change in Iraq later came back to haunt both father and son.

Yet Bush the elder left a valuable legacy: the use of diplomacy to the maximum, accompanied by the use of force to the minimum, in order to subdue evil.

George Bush the elder had one more mission to fulfill. He was — perhaps providentially — to preside over the disintegration of the Soviet empire. How he did it is worth recounting, and adds to his modest greatness.

The Fall of the Iron Curtain

In a famous speech while visiting the United States in March of 1946, Winston Churchill put his own, inimitable, semantic touch on Soviet communism. He referred to its totalitarian tendency to fence in its people, prohibiting escape at all its borders, as an "Iron Curtain." Half a century later, it fell to George H.W. Bush to preside over the dismantling of that awful, enormous veil of oppression.

Bush was not the architect of the victory of good over evil represented by atheistic Marxism. Yet he presided over it with such magnanimous grace that he deserves a lot of the credit.

Jon Meacham lists various factors in the ultimate collapse of this macabre, but oh-so-instinctively attractive philosophy, which still holds tiny countries like my own native Cuba and one, scary colossus (China). Here's Meacham in his own words:

> *The Cold War ended for many reasons. Among them were the aspirations of millions of ordinary people behind the Iron Curtain; the leadership of men such as Lech Walesa and Vaclav Havel and Pope John Paul II; the reforms of Mikhail Gorbachev; and the decades long resolve of the West, particularly the United States, to stand strong against Communism.*

The same could have been said by Meacham or any historian worth his salt as to the advance of civilization over the forces of barbarism and conquest by those who live by the sword, instead of their own hard work and collaborative enterprises that combine capital and labor into prosperity for all. The only difference is that it was not a "decades long resolve" but a centuries-long resolve.

Science and the Theory of God

More specifically, it was twenty centuries of Western resolve, buttressed by Western values.

Could humanity have evolved by itself, without the aid of the two great religious inspirations? Would a materialistic world, ruled by nations with no ideals of self-sacrifice, self-control of appetites, unrestrained dog-eat-dog subservience of women by men, children by adults, subjects by their rulers, have survived these last two thousand years, particularly if Nazi-type regimes had come to possess nuclear power and the capacity to launch intercontinental missiles?

Arthur Koestler grappled with the question, suggesting that "the schizophysiology inherent in man's nature" had to be cured soon. And he warned that if we "failed to find this cure, the old paranoid streak in man, combined with his new powers of destruction, must sooner or later lead to genosuicide."

Ironically, what Koestler thought was the *cause* of the problem is actually the *cure*.

I think the evidence indicates that we have found only one effective cure for humankind's tendency towards self-destruction: It is to reconnect with the Big Banger, to follow the better angels of our nature.

Historian Norman Davies describes how a fallen Nazi Germany had failed to nurture its "organic nature," and had to be reconnected to its soul, a task assigned to the poet T.S. Elliot and carried out by broadcasts.

Here's a passage that condenses the lessons of the poet's lectures:

> *'Culture is something that must grow. You cannot build a tree; you can only plant it, and care for it, and wait for it to mature...' And he stressed the special duty of men of letters. Above all, he stressed the centrality of the Christian tradition, which subsumes within itself the 'legacy of Greece, of Rome, and of Israel.'*

I leave to my reader to judge for himself and herself whether that legacy is, in all probability, the best, or the only, proven solution to the bad angels in our nature.

The answer is clear to me. Judeo-Christian explanations of human nature are the most consistent with the facts that indicate a deep flaw in our nature as well as a viable plan of redemption for both individuals and societies.

The fact that the messenger was a unique historical figure only enhances the believability of the scientific model. Could Jesus be understood in any context other than as a purveyor of salvation for a species seemingly condemned to selfishness, stupidity, venality and failure?

The answer to the question is in the next chapter.

Chapter VII

WHO WAS THE "SON OF MAN"

When he appeared, the soul felt its worth. In his name, oppression shall cease. From the lyrics for "O Holy Night" by Placide Cappeau

There are only two possibilities about the nature of the person that history has come to refer to as Jesus the "Christ," meaning the "anointed." Either he was made of flesh and blood, as all other members of our species, or he was some sort of superman: a god-man if you will.

The idea of a god-man is not particularly unique. Ancient mythology is replete with stories of human beings who were endowed with divine power. They were often a genetic cross between a god and a human – typically with the god being the male side of the human-divine sexual pairing.

(One suspects that the story-tellers, who were typically men, were enthralled by the female form; and that led them to assume the gods would be equally smitten and consequently willing to descend a rung or two in the hierarchy of beings merely for romantic purposes....)

As for the biblical writers, they curiously refer to Jesus as the "son of man," and only very infrequently as the "son of god." Given that they all ultimately concluded that he had divine powers, it seems almost as if they wanted to emphasize to their listeners that this man, who did and said such extraordinary things, was made of flesh and blood, just like them.

A very erudite thinker, Thich Nhat Hahn, in a most interesting book titled *Living Buddha, Living Christ*, who has studied recent, somewhat unorthodox documents that were allegedly written by Thomas the apostle, had this to say about the divinity of Jesus:

> *All Christians, while praying to God, address him as father. Of course Christ is unique. But who is not unique? Socrates, Mohammed, you, and I are all unique.*

Thich Nhat Hanh answers his own question as follows:

> *Because God the Son is made of the energy of the Holy Spirit, He is the door for us to enter the Kingdom of God.... The Buddha is also described as the door, a teacher who shows us the way in this life.*

Metaphors like the above, derived from Buddhist thought, are useful. Certainly, the fact that Jews, Christians, Muslims and Buddhists comprise the majority of the world's population lends credence to the idea that if we must reconnect or "relink" with God, a "door" or linking agent might make the process easier.

Admittedly, we cannot possibly resolve, with any kind of scientific certainty, the dilemma of Christ's alleged divinity. Whether he was or was not cannot be tested, 2,000 years after the human side of him passed away. Besides that fact, we are dealing here with a unique, historical phenomenon. Like creation itself, redemption (if it happened as stated in the bible) is inevitably a one-time experience. And, of course, it doesn't depend on us and cannot (by definition) be caused by us humans.

It has to be an act of the Big Banger, in the same way as we have concluded was the initial Big Bang, the subsequent inflationary moment (a/k/a "god's little workshop") and most likely both the beginning of organic life (biogenesis) and the beginning of the human species.

Science and the Theory of God

Those are four significant events that seem to involve an outside force or Big Banger.

So let's be frank and admit that there's no way we are going to empirically prove that a god acted on the planet earth around the time that Jesus was born, infusing special powers on this one human being, born of a human mother just like the rest of us.

But there are other scientific insights that we can bring to the analysis. Among those is the notion previously explored of a historical or sociological moment of "reverse entropy." What I mean is a moment in history when things changed so radically, when the collective psyche of society morphed so quickly and so unexpectedly, that one senses the appearance of brand new forces.

Clearly, the arrival of Jesus in human history generated a seismic movement akin to a sociological volcano of unearthly proportions. Culturally and sociologically, it represented such an important event that for two millennia the large majority of the civilized world uses Jesus's birth as the marker for the beginning of a new era of time.

Those things have already been discussed. Here what we are exploring is the psychological genius of Jesus, which doesn't necessarily imply infinite powers, but certainly sets him apart as a historical figure with no peer.

Jesus was unique for two substantive reasons: (1) his insistence on forgiveness as a normative rule of human conduct, and (2) his diametrical reversal of the instinctive human hierarchy, which requires that the strong be at the top of the heap and the weak and small at the very bottom.

But there is also a uniqueness of style in Jesus that begs consideration. There is a totally unprecedented and inimitable charisma, a stage presence coupled with a grasp of human tendencies and weaknesses, an ability to solve circumstantial impasses that defies comparison and begs for an otherworldly explanation.

To me, the best example of that charisma and that political astuteness was the scene with the prostitute who had been exposed and for whom the agreed-upon punishment was death by stoning.

Xavier L. Suarez

Jesus as the Politician

I only know of one instance, in all of the New Testament, in which Jesus was seen to write something that no one could decipher. The scene was as follows.

There was a mob, and Jesus reportedly passed by as the mob was getting ready to stone a prostitute to the death. One can imagine the most brash in the mob yelling out the accusations, with details as to time and names of eyewitnesses. One can imagine the most violent, or those with some personal guilt of their own (perhaps a jilted lover of the very accused, or one of her former clients) getting ready to hurl the biggest brick that could be found within reach.

What changed the momentum of the macabre scene was not Jesus's appearance on the scene. From biblical accounts, what changed the momentum was the fact that the Jewish elders decided to put Jesus to the test. One can imagine them presiding over the execution-by-stoning, perhaps in a group of three or five, standing on one side. One can imagine that the executioners were waiting for a signal from them that the prostitute was, indeed, guilty of the crime and that they were authorized to carry out the legally prescribed sentence.

Then, all of a sudden, the group of legal scholars approaches this bearded man, of rough hands, and broad shoulders, with a gaze that bespeaks of authority, kindness, compassion, and every other quality we ascribe to both the most alpha of alpha males and the most sensitive of sensitive men. (Think of Robin Williams inside the body of Charleton Heston.)

As a politician, it is never a good thing when you are put on the spot. If there is a public gathering, you are careful not to walk into the trenches, lest you be identified with the multitude; if the multitude is more like a mob, you are careful not to get even close to the action. When people are angry, rebellious and prone to violence, a leader has to be very, very careful not to weigh in at just the wrong moment.

I have experienced many of these circumstances, so I can appreciate what Jesus did. In 1989, I walked into a civil disturbance, following the unexplained and somewhat unexplainable shooting of a black motorcyclist in an area of Miami where the population is mostly black, mostly poor, and mostly neglected by the powers-that-be.

Science and the Theory of God

Any political consultant would have counseled me not to do what I did: For political leaders, walking into an angry mob is always the wrong thing to do.

The right thing to do is to send your aides, your "advance men" into the scene. They reconnoiter the scene, separate the troublemakers from the real community leaders, make all kinds of promises, and act as intermediaries between the aroused rabble and the eminent leader.

The last thing most leaders do is just walk into a confrontational scene. *Most politicians don't like to rock the boat, let alone walk unescorted into a boat that is already rocking.* In this case, Jesus walked into a nasty confrontation, of the kind that was both macabre and titillating. We can barely imagine the kinds of human emotions being poured out in the stoning of a prostitute: cruelty, vengeance, misplaced righteousness, imposition of sexual superiority, physical exertion, even a sort of deranged athleticism on the part of those hurling the stones.

These were times in which the rulers of Rome and the entire Mediterranean Sea offered their subjects spectacles that would shock us. In those times, a typical Roman circus involved the gore and titillation of a gladiator going against a lion. But even that lacked the sexual content and religious justification of stoning a prostitute, which allowed the observer to enjoy the misery of the suffering victim, while feeling righteous about the whole ordeal.

Jesus was the ultimate reformer. He did not care one whit about convention. He did not accept authority when it was improperly applied. Nowadays, we would say he was into "disruptive" methodology.

Let's observe how Jesus handled a situation that would have scared away almost every politician I know.

Jesus Confronts the Legal Scholars

On being approached by the rabbis, Jesus probably wondered, "why me, and why now?" He understood that he had walked into a tinderbox. On the other hand, he also perceived that this was the ultimate "teaching moment." These wise old men were putting him in the position of being the judge not only of the woman's punishment, but of the validity of the law that allowed men to kill women who were unfaithful, but did not give any corresponding right to the women.

As if directing a movie, Jesus takes the opening given to him and proceeds to put a wholly unexpected stop to the action. He doesn't scratch his head; he doesn't stammer; he doesn't show fear; he doesn't skip a beat. But he does buy time. How does he buy time to think? By bending down, squatting and calmly writing some sort of message on the dirt.

And what did the message say? Well, the biblical writers don't seem to recall. Or maybe it was not a message at all.

I have heard some of the great homilists (accomplished theologians in every case) describe the scene and offer a possible interpretation of the "message" written on the dirt. The most common theory is that he read the heart of the rabbis and/or of the rabble and listed some of their own sins, as a way to shame them into retreat.

But if he wrote something legible, why wouldn't the disciples have recorded it for history? And if it wasn't legible, or had been erased by the mob, why would they not have questioned him afterwards, so as to complete the lesson of this important teaching moment?

My theory is much simpler. My theory is that he was doodling – writing whatever symbol came into his mind, as a way to gain time and to keep control of the scene. He understood that the longer the rabbis had to wait for his answer, the more legitimacy they gave to his pronouncement. He might have also hoped that the hotheads holding the biggest stones would cool their heels, retreat to the background, and wait for the intellectual confrontation to play out.

It was the most brilliant political handling of being on the "hot seat" since Solomon had been asked to decide between two women who claimed they each were the mother of a child. Much later in history, there is a similar scene involving George Washington, to which I shall shortly return.

But first a bit of reflection. Let's assume that Jesus had a direct line to what Deepak Chopra calls the universal intelligence, or the entity which theoretical physicists refer to as the Big Banger, or the Greeks call the "logos" or that most of us refer to as "god." If Jesus had a lifeline to rely on, and if that lifeline was not always activated, it might take a minute or two to connect and ponder the advice received from afar. Even the fastest computers, which send messages at 186,000 miles per second (the speed of both light and electrons), take a little while to turn on and "load" one's email messages.

Jesus was human, so it took him a while to come up with a solution, to be ready for the showdown with the teachers of the law. He begins by standing up to his full height. In a clear voice, he pronounces his verdict: "Let he who has no sin cast the first stone." You can sense the reaction of the rabble and the frustration of the rabbis. The rabble are thinking: "Well, under that standard, we certainly cannot take the initiative, or we can be accused of thinking we are better than our religious leaders; let's see what they do."

The rabbis are thinking, "Well, we are not built for this kind of thing. We don't have the arm strength, or the sheer crudeness that it takes to stone someone to death. Plus, what if this so-called prophet can see through our righteousness to our soul, and be shocked by what he sees? The man does not seem flustered by our trickery, or afraid of the mob. Hell, we placed him in command of the scene, and now we are made to look like fools."

And so they walked away, defeated intellectually and spiritually. They would live to fight another day.

Is Jesus's performance here worthy of a superman? Of a divinity?

Before we answer that, let us look at some other facets of what we know about his life. One incident happened at such a young age that it was either a fabrication or a manifestation of a child prodigy

Jesus the Child

There are two glimpses of Jesus before he made his first public appearance. One was in a visit to the temple, when he was about twelve years old.

We are told that he got lost, perhaps when he was not allowed to participate fully in purification rites involving only adults. He reportedly worked his way into a discussion with the teachers of the law, and was apparently holding his own with the very learned professors when his parents found him. One would guess that he worked himself into the discussion by perhaps posing an innocent question, which contained within it enough scholarship to amaze the learned men assembled there.

I would guess that the first question he posed was followed by another seemingly innocent question, but containing even more wisdom in the premise embedded within. Soon, the child was seen participating in both questions and answers, evincing biblical scholarship not just beyond his years, but beyond his era.

That in itself is extraordinary, but not as strange as what happened next, when his parents scolded him for making them think that he was lost (and perhaps kidnapped or otherwise in danger). "Don't you know we were concerned that something had happened to you?" they asked.

His answer gives the first indication that he was conscious of having a different mission – perhaps conscious of inner insights about his otherworldly persona. He said, simply and bluntly: "Don't you know I have to go about my father's business?"

We all have ideas of grandeur. Some of us have had them since we were adolescents. But Jesus seemed to have a better view of his own transcendence than any adolescent I know or have read about. Moreover, he seems to have had a professorial grasp of scriptures, to a degree that enthralled the professionals.

That kind of virtuoso performance by a youngster has been seen in the arts – particularly in music; Mozart is the most-cited example. But music and art have a high intuitive component, whereas scriptural scholarship must be acquired through extensive reading. Could Jesus have mastered scriptural scholarship as a twelve-year old?

It makes one wonder.

Jesus as Wedding-Planner

As a natural theologian, I don't assume the veracity of any miracles. I cannot prove, to any degree of scientific certainty, that Jesus was able to turn water into wine. But I do marvel at the psychological wisdom of this incident, which is centered around a festive event: a wedding.

The first reported miracle by Jesus was unusual in two respects: (1)

it happened at a party and (2) it involved an alcoholic beverage. Each feature is interesting.

I recently heard a preacher remark that any follower of Jesus that doesn't like parties is not a very good follower of Jesus. Think about that for a second. What the homilist was saying goes against the grain of what many preachers have been saying for thousands of years, which is that religion is not supposed to be fun.

From the Greek Stoics to the Catholic ascetic orders (and, in an exaggerated distortion of Christian theology, the Manicheans), there have been many proponents of the idea that to be religious one must begin by discarding pleasure.

The great fiction writer, Tom Wolfe, wrote a novel entitled "Bonfire of the Vanities." It depicts a character who had a decadent lifestyle, full of every human pleasure, and falls through a rather narrow and accidental crack that tears away every vestige of an honorable life. Wolfe's protagonist goes from riches to rags in minutes.

The title of the novel is an allusion to a historical event. It was an actual burning of books by a convinced Christian who felt that all books (other than the bible itself) were works of the devil, for the simple reason that they excited our pleasure nodes and thus distracted us from the essential task of being a good human being. His name was Savonarola, and his core belief was that life on this earth was supposed to be one big drag.

Christianity, to Savonarola, was no fun at all. To be followers of Jesus, to earn redemption from sin, to be united with God in heaven forever, one had to renounce all earthly pleasures (vanities) and become an ascetic monk.

It was like being St. Francis of Assisi, minus the joy.

The Savonarola episode evokes the "Valley of Tears" model of Christianity. It doesn't work for me and it probably doesn't work for most of humankind. It also doesn't fit the personality of Jesus.

That personality is not only eclectic but unique.

The Parable of the Talents

For me personally. the most mesmerizing, the most captivating, and the most haunting of Jesus's pronouncements was the parable of the talents. In this story, Jesus seeks to be the Pied Piper of an obligatory form of altruism that some writers have called "noblesse oblige." Loosely translated from the French it means something akin to a combination of what Plato said about those who were "golden" people from birth with the Jewish and Christian idea of "giving back to society."

This goes back to the point previously made: perhaps redemption is not one-size-fits-all. Perhaps some members of our species are born with more potential for mischief, while others are born with more potential for good. I know this: From an early age, I was conscious of having certain gifts. I was also conscious that with those gifts came certain responsibilities. It wasn't just the biblical narrative, but I have to admit that the bible story affected me enormously.

It was the parable of the talents.

The biblical parable is simple: the master gives his servants different amounts to invest while he travels. There are some who receive large "talents" to invest – to manage so as to obtain substantial, proportionate returns. Then there is one who gets just a couple of meager coins; he becomes fearful of losing even that meager amount and decides to bury it, lest someone steal it.

The master is angered by the shy, fearful servant. He excoriates him, while lavishing praise on the more talented, risk-taking servants. Then he pronounces the moral of the story: "From him to whom much is given, much is expected."

Those words have haunted me all my life. Although I was the ninth child in a family of fourteen, I have never lacked the essentials for a contented, even privileged life. As a child, I lived in a large, modern house across the street from the campus of an Augustinian university (Villanueva) where my dad was the dean of the engineering school.

All my life I managed to get into the best private schools and excel not only in academics, but in sports. Math, languages and science came easy to me. I was elected mayor of Miami at 36, reelected twice, worked for the best law firms in town and now enjoy a second life as a politician, being a county commissioner in one of the world's most powerful county governments. My marriage has lasted four decades and has produced four children and eight, gorgeous grand-children.

Except for a few sports injuries, I have no physical ailments and no economic concerns. I have never spent a night in a hospital.

There is not one iota of doubt in my mind that what Jesus said applies to me: that a whole lot is expected of me. I am absolutely, categorically sure of that. I suppose I would think that way, even if Jesus had not said it in such vivid terms and even if preachers and writers and movie-makers had not repeated it and illustrated it for the last twenty centuries.

But that is pure conceit on my part. The reality is that if Jesus had not lived, and if his teachings (based as they are on those of Abraham, Moses and the rest of the biblical prophets) had not permeated to every successful modern society, I would probably not be haunted by the need to "give back" at all. To the contrary, I would probably enjoy my riches, flaunt my talents, step around the misery of others, and reach for the next cup of pleasure as if I were entitled to my gifts – rather than blessed by them.

What life lessons would a relatively young Jewish carpenter's son have in his background that would empower him to come up with such vivid insights, wrapped inside such amazing oratorical embellishments?

Based on all the available evidence – none. And other vignettes are even more powerful, as we shall see.

Jesus as the Psychologist

Science has very little to contradict about the historical Jesus – the one described in three almost identical accounts (called the "synoptic" Gospels) and one rather mystical account by someone we have come to call St. John. History seems to support most of the Jesus story, though it is admittedly sketchy, consisting pretty much of one Jewish writer, who later converted.

That historian is known by the name of Titus Flavius Josephus, though his birth name was Joseph Ben Matityahu.

Later historians and anecdotal evidence confirms that there was a historical character named Jesus and that he not only said some extraordinary things, but also that his pronouncements and deeds quickly and radically changed the course of history.

O.K., but so did Marx and Mao. One salient difference is that Jesus claimed to have divine powers. And he thoroughly convinced a group of twelve close followers that he was the Messiah, the anointed one who would usher in a new era of human relations.

It took a while for his followers to understand just what was being proposed by this Messiah, who didn't act like the Messiah they were expecting. But they became convinced of his message, and were willing to die to proclaim it. Ultimately, it took the better part of three centuries for his followers to convince the emperor, and the emperor brought with him the largest empire of the times.

Ultimately Jesus's followers convinced the entire Western world of their vision.

Mohammed came a couple of centuries after that, and he also claimed divine inspiration. Today he claims just as many followers as Jesus.

But Mohammed did not change the world like Jesus did. Some of his more radical followers have shown an uncanny ability to wage war, and to subdue those who are physically weaker (i.e., women); but, as a group, his followers have not distinguished themselves anywhere near as much as the followers of Jesus.

In conclusion, the impact of Jesus, as a historical figure, is unique. No reasonable person questions that. No historian worth his salt questions it. No fair-minded person, looking at the historical facts, questions it.

But was Jesus divine? In the next section, we assume that he was – for purposes of discussion. Think of it as an intellectual or procedural exercise – like in parliamentary procedure when someone makes a motion and you second it, but only so it's "on the table" and can be discussed.

You don't necessarily have to vote for it. Just consider it and see where it leads you.

Parenthetically, it is not an outlandish assumption to assume that someone who changed history as much as Jesus did, possessed an element of "superman" in him. The man changed not only humanity's timetable, but also humanity's way of thinking about revenge (not allowed), the supremacy of physical power ("meekness" is preferable), tribalism (the Good Samaritan is as good as a Jew), and, in general, the

Science and the Theory of God

entire notion of equality of all, from the most powerful to the lowest inhabitant of a leper colony.

Think about it. When Winston Churchill, prompted by Gandhi, accepted the notion that India would gain independence, he struggled with the reality of a nation in which 60 million people were considered not just inferior, but "untouchable." He struggled with the reality that for many civilizations, the idea that humans are all equal in dignity is simply unacceptable.

Churchill was convinced that sooner or later, all nations were bound to enter the community of free nations, all would accept the notion of the inherent equality of all humankind. He set out to accomplish India's independence, while also understanding the reality that the Muslims who mostly inhabited Pakistan would never consider the Hindus who mostly inhabited India, their equals. (And vice-versa.)

So Churchill crafted a policy of "separate but equal" and carefully divided the nation into two smaller nations, each with its own religion. (The book to read is *Freedom at Midnight,* by Dominique Lapierre and Larry Collins.)

Americans of all races and ethnicities profess one bible. The small minority who consider themselves atheists or agnostics accept, almost as second nature, the more enlightened lessons of that same book. They have managed to stay together as one polyglot nation precisely because they are united in Judeo-Christian values, whether because of their faith or because of their reason.

In sum, Jesus upended the old order like never before in history. And, if my sense of history, science and logic serves me right, like never again. There is no need for any new prophet. Jesus embraced all the prophets that came before him, fine-tuned all their teachings, and gave us a refined understanding of the Decalogue that might just propel humanity to a just and lasting peace.

So, let's see where the assumption of a God-man leads us.

What Sense a Half-Man, Half-God?

In this section, we assume (even though we can't prove it), that Jesus was part man and part God.

How would that work? Would the divine will, and its concomitant clarity of understanding, always rule the decision-making by Jesus? Or would there be a mish-mash of wills, a clash of understandings, the kind of tension that we humans experience between fear and resolve, selfishness and generosity, short-term goals and long-term vision?

Biblical accounts seem to support the notion that Jesus had an inner tension between his allegedly divine nature and his human nature. We see that in the scenes that depict a man torn between what he sees as his mission (including being crucified) and his human wellbeing.

We see that in scenes that depict a man tempted to fly away from pain and hunger and humiliation and enjoy the best that human society has to offer: a kingly banquet with lesser humans prostrate before him, ready to do his bidding.

Here is a man, who has been fasting for longer than a month, and he sees visions of being fed by legions of adoring angels. Here is a man, who senses that he is about to be reduced in his understandable aspirations to be fully human, to share the life of the average citizen: eat, sleep, spend quiet time with family and attend an occasional party with friends; have a cup of wine; celebrate the life of a carpenter's son, in an area of the world that was stable.

As a Galilean, Jesus would not be at the top rungs of the Roman Empire, but he would not be a slave, or an outcast. He would not be in constant fear of starving, or being mistreated by the authorities. Jesus was a proud member of a strong religious community; he was brought up strictly according to its practices. He could not be king, or senator, or a member of the ruling class, either within the Hebrew society or the Roman occupation force.

But he could, and certainly would, yearn for a decent life as a craftsman and a Jew, in a land where Jews had ample room to live freely and worship freely – though not to govern. (Which is why the average Jew yearned for the Messiah to liberate them from Roman rule.)

But Jesus was a lot more than a prophetic member of a people who collectively felt chosen by God. His words and his deeds are those of someone with something rare inside.

Whatever one thinks of him, he was a unique, dramatic figure in history. If not true, the story of Jesus is still the greatest story ever told. That's impossible to deny.

But is it more than that?

The Christ of the Gospel

The Christian vision of Jesus is captivating. Here is a man, who also happens to be a God, experiencing human emotions as diverse as adoration one day and crucifixion the next. How exquisite! What drama the life of Jesus....

But did it have to be that way? Or is this all fluff? Does it stand the test of history? Is this theory consistent with science?

By this I don't mean that Jesus's divinity (or his alleged miraculous powers) can be proven, as some people try to do with the Shroud of Turin or other supposedly miraculous amulets or events. What I mean is: Does Jesus meet a rational test of believability as a historical character?

Undoubtedly, the historical view of Jesus presents a classic example of what we mean when we say that someone is "bigger than life." But how much "bigger than life" was he?

Mahatma ("Great Soul") Gandhi was bigger than life. This thin little man, whom Churchill once called a "half-naked fakir," changed the course of history for a whole continent.

Martin Luther King, with his amazing eloquence, mobilized legions of non-violent protesters and changed the laws of the American South – a huge and powerful section of the country that had, less than a century before, fought fiercely to preserve slavery.

His "I have a dream" speech has reverberated over the years and become a powerful anthem to the equality of the races.

Abraham Lincoln waxed poetic at Gettysburg; this memorable performance brings to life a special, theatrical moment, with an unlikely protagonist. Lincoln was emblematic of a rough-hewn, self-taught, unattractive yet amazingly charismatic man who united a nation when it could easily have been torn asunder. Would anyone deny that he was "bigger than life?" (The book to read is Doris Kearns Goodwin's *Team of Rivals.*)

Joan of Arc had that quality. Once again, we turn to Churchill for a description of her sheer charisma:

> *Joan was a being so uplifted from the ordinary run of mankind that she finds no equal in a thousand years. She embodied the natural goodness and valour of the human race in unexampled perfection. Unconquerable courage, infinite compassion, the virtue of the simple, the wisdom of the just, shone forth in her. She glorifies as she freed the soil from which she sprang.*

Joan of Arc and George Washington are two of a kind. They came into the scene as people with a vision that was unequaled, a knack for leadership that was astounding, and a firmness of purpose that bespeaks of some inner force that is simply without precedent.

France would not have been the nation that ultimately gave birth to the Renaissance without Joan of Arc. You could say the same thing about King Henri of Navarre, who solved the religious dilemma by aligning himself with the majority, at just the right moment. But Henri of Navarre was a full-fledged king, of noble ancestry. Joan of Arc was the female daughter of peasants, at a time when peasants (and women) were not even literate. She united a nation on pure instinct.

If I had to compare anyone to Jesus in terms of unprecedented, unalloyed, inexplicable talent for leadership, if would undoubtedly be Joan of Arc. But for sheer stage presence, no one comes close to George Washington.

The Bigger than Life Figure of George Washington

I have read a lot of history – though I am not a historian. Unlike Herodotus and Titus Flavius Josephus and Thucydides and the Gospel writers and the Middle Age historians, I have the benefit of the Internet. Historical research is accessible by requesting a book from the public library or punching a key on this machine.

When I do, I get the benefit of reading Gibbons, William Manchester, Robert Caro, David McCullough, Doris Kearns Goodwin, Jon Meacham, and a host of modern historians.

In my readings, there are no "bigger than life" figures that measure up to George Washington, as far as stage presence. (I should say in modern history; if proven to be fact, the David-versus-Goliath confrontation had to be the number-one scene in history for not only stage presence, but performance under stress....)

In modern history, the apex of performances under stress – of sheer stage presence and mastery of difficult circumstances – was reached by revolutionary war general George Washington. Historians tell us he was facing a mutinous mob of officers in Philadelphia. To me, it was *sui generis* in modern history, analogous only to Joan of Arc at Orleans.

Let's see what that was like and compare the two scenes, for sheer sense of the dramatic and intellectual *gravitas*.

Washington with the Mutinous Officers

There is a scene in George Washington's life that many biographers consider his finest moment – though not in the strictly military sense. One biographer, Ron Chernow, calls it the most famous "*coup de theatre* of his career." It happened towards the end of the war against England. The stage was set when a large group of ranking officers were ready to mutiny in protest for their lack of wages and other essentials needed to survive.

General Washington decided to attend their meeting, knowing that anyone he sent might not make it out of there alive. The meeting was held in the "Temple," a 40-by-70 foot building at the camp.

After General Gates opened the meeting, Washington entered the building, to everyone's surprise. He asked to speak to the officers, and the stunned Gates relinquished the floor. Washington could tell by the faces of his officers, who had not been paid for quite some time, that they were quite angry and did not show the respect or deference to their commander as they had in the past in the past.

Washington then gave a short but impassioned speech, now known as the "Newburgh Address." For a modern leader, this was the classic no-win situation. Some sort of prop, some special performance, some particularly poignant inspiration was *de rigueur*.

If you didn't perform, you could end up either as the head of a non-existent army or as the headless, former general who was killed by his own officers.

George Washington, on the spur of the moment, decided to use a prop.

The prop he used was a letter from a member of Congress, which he decided to read to the officers. For a few seconds, he fumbled with it without speaking. The fumbling may have been accidental, but what happened then was intentional. Adding to the dramatic moment, Washington proceeded to take a pair of reading glasses from his pocket. (It should be noted that they were new; few of the men had seen him wear them.)

He then said: "Gentlemen, you will permit me to put on my spectacles, for I have not only grown gray but almost blind in the service of my country." This was a poignant reminder that Washington had not only sacrificed a great deal for the Revolution, but had done it for as long as any of his men. Historians tell us that many of those present were moved to tears.

History records this moment as one in which a member of our species – a particularly gifted member – shone with a special brightness. He seemed "bigger than life" in the sense that he rose above the biological limitations of our species.

David slew Goliath. Moses parted the Red Sea. Jesus tamed the Roman Empire. Joan of Arc united France. Washington stood down his mutinous troops and defeated the British Empire. Wilbeforce made slavery illegal in all British colonies. Gandhi dismembered the British Empire. Churchill defeated Nazism and put an end to the Holocaust.

Those are eight, eventful, historical moments. Those are eight figures who stand out as being "bigger than life." But only one seems to exceed all bounds of reason, all expectations of greatness, all measures of what our species can accomplish.

By any quantitative or qualitative measure, Jesus was history's greatest over-achiever. That is not a theological assertion. That is a scientific observation.

And I am willing to back it up. I consider myself a pretty accomplished politician. I have been mayor of a major city. I am currently a commissioner of a county that is one of the largest and most powerful in the world.

I have been praised for walking into a major racial disturbance

Science and the Theory of God

(1989). I have been selected by Esquire Magazine as the best local official in the United States and by Howard Gardner (in the *Harvard Project on Multiple Intelligences*) as an example of "high interpersonal intelligence."

But I could never have even begun to fathom the marvelous, spontaneous, unrehearsed stage presence of Jesus as he dealt with a bloodthirsty mob combined with a self-righteous group of intellectual elites.

I marvel at the composure he showed, the wits he marshalled and the spectacular, unexpected, historically unprecedented *denouement* that ended the stand-off between upstart preacher and the establishment.

The scene was the stoning of a prostitute.

If it happened as we are told by the Gospel writers, it was an extraordinary moment in history. I don't know and have not read of anyone who could have pulled off what Jesus did, with the witty retort and the dramatic flair that he invoked.

Not even Gandhi, with all his moral power. Well, maybe Gandhi or John F. Kennedy, or Martin Luther King. But those men were all accomplished leaders, with post-graduate degrees and worldwide name recognition. They had legions of followers, to back up a bold, spontaneous roadside intervention.

Jesus was a carpenter's son surrounded by a band of twelve scared disciples.

As I said at the beginning of this book, the three great philosophers (my "A-Team") did not have the benefit of modern psychology. Neither did Jesus.

And yet Jesus showed extraordinary psychological insight.

Jesus as Clinical Psychologist

I have previously mentioned the remarkable wisdom of Jesus's advice on the issue of judging others. Now I want to expand on that theme, with a little more dissection of what he actually said.

Notice the metaphor he used: "Why do you dwell on the speck in your neighbor's eye, while ignoring the beam in your own eye?"

What Jesus started, with this admonition, was subsequently complemented very nicely by the special grant he gave his disciples when he said, "whose sins you forgive, are forgiven." This is the other side of a very powerful coin, whose front side is self-examination and whose back side is guaranteed forgiveness.

There is no greater, more exhilarating prescription for cleansing a troubled psyche than the combination of critical introspection with guaranteed forgiveness. Even if you don't add the Catholic element of special (supernatural) graces accompanying confession, the combination of these two doctrinal prescriptions is compelling.

As with the publication of the Ten Commandments, which constitute a fundamental code of conduct, the double lesson mentioned above is nothing less than a bedrock code of psychological self-correction, self-improvement, and self-healing that is unequalled in the entire science of psychology, when serious psychotic disorders are excluded from the analysis.

In effect, Jesus was anticipating, with the first admonition, the modern concept of "projection." He was ahead of his time by 1900 years, when he noted that we humans project our own failings on others: We are inclined to notice in others the very faults that we have. And, of course, Jesus did it with a speaker's hyperbolae: the **speck** we notice in others is no less than a **beam** in ours. We are totally blinded by that beam, while pretending to root out a little speck in our neighbor's eye.

It is the first time in recorded history that anyone suggested we should judge ourselves before we judge others. That was followed by a commission given to those who would have spiritual authority over the multitudes: They are told that if the sinner (which is all of humanity, by definition) objectively and humbly examines his conscience and admits his failings, he/she is forgiven.

But he/she must first open his/her heart to another human being, in what has come to be called a "confession."

The Secular Value of Confession

The Catholic ritual of confession is interesting enough to Hollywood that it seems to have a part in something like one in ten movies about the Mafia (and not much less than that in movies featuring Irish-Catholic politicians). The idea that a goodly portion of a billion people find forgiveness by baring their soul to some guy wearing a cassock and hiding behind an opaque partition is clearly enthralling to film-makers.

It sounds like something out of the Middle Ages, when penitents would proclaim publicly their transgressions and were then required to cover themselves with sackcloth and wear ashes on their foreheads for weeks on end. But it actually has strong secular, scientific value.

If you combine the idea of confessing sins to a third party with the idea that one must try to love others as much as oneself, you get the following scientifically compelling steps towards a healthier, more socially valuable, more caring personality:

Step 1: The person is expected to carry out a thorough self-examination, in which the mandate is to objectively analyze the failings in the immediate past life. The bar is set high: the mandate is to love others as much as yourself. Objectivity is furthered by other elements of the three-step process.

Step 2. In order to expect forgiveness, the person is required to affirm the objective of improving his or her personality, being more loving and caring for others. If that commitment is lacking, the third component has no expected value.

Step 3. If the above steps are followed in good faith, the person is guaranteed forgiveness, erasure of all guilt, and renewed joinder with the community of faithful.

Note the scientifically sound aspects of this practice: The person is required to attempt self-examination, i.e., evaluating one's conduct towards others as compared to how one wants to be treated. The mandate is simple: "Love your neighbor as much as you love yourself." (Makes one wonder if Jesus had anticipated the bestselling book, *I'm OK; You're OK*, written in 1969 by Thomas Anthony Harris.)

Comparing one's self-esteem to one's esteem for others contains an important psychological insight. Basing one's conduct on the love of self assumes that a person has high self-esteem, which is an important component of psychic health. That is followed by some harsh soul-searching, of the kind we all should do from time to time.

Secondly, the person is required to tell another human being to what extent he or she has failed in being a totally caring, loving person. Telling others what is in our hearts is a very important healing practice for most of us. If the person on the other side of the confessional partition is mature, somewhat acquainted with the human condition and himself a practitioner of self-denial and self-correction, the feedback received is often akin to the best that can be given by a psychologist.

And it's free. There is no charge for confession.

If Jesus only knew how much he contributed to the psychological health of his followers with his prescription for self-correction, he would no doubt be surprised. It's hard to believe he didn't see the future at least a little bit, as we shall see when we examine in some depth the Sermon on the Mount.

Could a mere man, a carpenter's son born in Nazareth of a poor family, with little if any formal education, have concocted such a coherent, holistic prescription for psychological health as what I have described above?

I doubt it.

I realize, further, that it would take a humongous computer simulation to program what the world would look like without the aforementioned teachings of Moses and Jesus. As big and powerful as that computer would need to be, imagine one that could take into account the impact of the Sermon on the Mount.

The Sermon that Changed the Civilized World

Reportedly, Jesus preached his greatest and most enduring sermon on some hill or promontory. It stands, with Moses's revelation of the Ten Commandments, as one of the two most famous speeches in history.

In one little speech, Jesus praised eight kinds of people: (1) those who are "poor in spirit," (2) those who "mourn" some sort of loss, (3) those who are "meek," (4) those who "hunger and thirst for righteousness," (5) those who are "merciful," (6) those who are "pure in heart" (7) those who are "peacemakers" and (8) those who are "persecuted for righteousness sake."

Taken together, these admonitions essentially turn upside down the entire value system of what came before Jesus. Strength, violence, superiority, victory in the field of battle are denigrated, while caring, compassion and meekness in the face of persecution are glorified.

The scale of values was anathema to the philosopher who, unwittingly, paved the way for what was perhaps the most malevolent of all modern political movements: Nazism. I have already explained that his name was Friederich Nietszche and that he brought to the twentieth century the entire, distorted, malevolent edifice that justified racism, sexism and full-fledged oppression of the weak.

Nietszche objected strenuously to the values preached by Jesus. He thought that they fostered a "slave mentality," whereas we should embrace a "master" mentality.

His thinking gave intellectual support to a philosophy known as Social Darwinism; it fed the impulse of a very powerful, very technologically advanced people.

It was adopted by Adolph Hitler.

It says here that had it not been for the Judeo-Christian value system, ingrained as it was in England and the United States, Hitler would have vanquished Europe and either neutralized or conquered North America.

In many ways, the fate of not only Western but Eastern civilization hung in the balance. Ultimately, the Beatitudes won and Hitler's Master Race lost.

The Twenty-First Century Without Jesus

Winston Churchill and Franklin Roosevelt, plus the power of American industry, won the Second World War. Ronald Reagan and Pope John Paul II and Lech Walesa, aided again by the power of American industry (this time in the form of computers that could strike down Soviet missiles in space), won the Cold War.

Many people think the world could not survive a third world war. With the collapse of the Soviet Union and the advent of a capitalist China, with strong economic ties to what used to be called the "Free World" and now encompasses pretty much all nations except a couple of hardline, communist ones (Cuba, North Korea) and a couple of radical, Islamic ones (Iran, ISIL), the likelihood of a third world war has diminished.

I should add that the possibility that a third world war will have to be waged with China is the topic of a book by Harvard's Graham Allison. It was published in 2017, just as I was finishing this book. To Allison, the question is how to avoid "Thucydides's Trap," which is the scenario that occurs "when a rising power threatens to displace a ruling power."

Allison analyzes geopolitics from the standpoint of secularism. His recommendations, as to avoiding war with China, are all founded on communications, diplomacy, and the power of the human intellect to anticipate future events, if the present is analyzed dispassionately.

There is nary a mention of values, or of the impact of Judeo-Christian traditions or philosophy, as they have shaped modern civilization and as described in the preceding chapters.

If he and I had an open discussion, here's what I would ask him: Would the current trend towards normalcy, including a world council (the United Nations), a world court (the International Court of Justice) and a worldwide trade agreement (GATT) have ever come to pass without Abraham, Moses and Jesus?

I would love to expand the mythical Harvard forum, with which I introduced this book, to include a debate between Allison and myself on that question. (I would, of course, have some handy historians like Dawson and Meacham to help me, plus the novelist Tom Wolfe, who has opined on the importance of religious values to modern society.)

The question of whether our species would have destroyed itself by now without the Judeo-Christian influence has never been posed in this way – at least to my knowledge. I have never, in all my readings, seen an academic debate on it. I would like to propose it, as the concluding analysis of my modest attempt at determining how the mind of God works.

Intuitively, I have no doubt whatsoever. Had it not been for Judeo-Christianity, there would be no civilization today. Certainly not of the kind we enjoy – invested and ingrained as it is with the concept of equality, forgiveness and love even towards our perceived enemies.

Whether there is, in fact, an afterlife awaiting is another question – though related. God, if he/she exists, does not of necessity have to provide room in the eternal halls for any created species – even one that he/she has seen fit to provide with a free will and then provide with a way to reconnect that will to its full potential.

Scientifically speaking, all we have proven, with a fairly high degree of well-reasoned probability, is that the God theory is consistent with the various natural and social sciences.

Proving there is an infinite heaven is a stretch. But it does make sense, as we shall see. It resonates not only with the mind, but particularly with the heart.

Chapter VIII

THE KINGDOM OF HEAVEN

We are one, after all, you and I, together we suffer, together exist and forever will recreate each other. Pierre Teilhard de Chardin

One of the tragedies of the Christian era is that the followers of Jesus have not described heaven nearly as well as they have described hell.

Even Purgatory has gotten more ink than heaven. I have previously described C. S. Lewis' vision of Purgatory in vivid, believable terms that smacked of something real and tangible. But Lewis fails to give us a memorable vision of what heaven could be like – other than to say it is an eternal dance of some sort.

For most believers, heaven is defined more as the antithesis of hell than as something desirable in itself.

In one sense that is not surprising. How can any of us describe a place we have never visited? How can we begin to even decipher a hoped-for state of being that presumably starts when we die, continues for eons of time, and involves ethereal figures who have "resurrected" after dying?

Yet how can we not? How can we suggest to people that the Judeo-Christian equation of salvation is desirable, to the point of constant self-denial and even occasionally heroism in the face of death itself, if there is no substantial reward at the end of the rainbow?

There are two serious problems with the conventional, theological view. One is that while humans yearn for eternity, we have scant capacity to understand it. That particular problem can be surpassed, if not solved, by speculative analysis. We can extrapolate from our current appetites to a more complete form of joy. Whatever pleasant or joyful experiences we now have, we can conjecture exponential lines of increasing bliss and see where they take us.

In the final section of this chapter, I engage in precisely that kind of speculation. But a more troubling existential issue confronts me, and keeps me awake as I try to find satisfactory conclusions to this little voyage through science on the way to God.

What to Make of Humans Who Have No Yearning for Eternity

I have thirteen brothers and sisters, so that makes 14 siblings, counting me.

Of the fourteen, one was a nun. Towards the end of what appeared as a generally happy, fulfilled life, she seemed frustrated by the narrow range of social activities that her convent life offered.

The angst that we noticed, and that she conveyed more than once to me (almost as a legal consultation), was related to the inability to visit with older, non-religious Hispanics in the area where she had previously been stationed, which was the rather festive city of New Orleans. The forced transfer from a place where she had some lively socializing to a convent that was basically an infirmary for older women (many of them affected by various degrees of dementia) was traumatic.

For a while she rebelled. Then a fortuitous thing happened: Her own mental infirmity (diagnosed as Alzheimer's) began to take hold of her, with the result that her rebellion subsided to a mild form of restlessness. It was as if her inner demons had been overcome by the angelic music and the cadence of prayer in her religious enclave.

That steady-state condition, I should add, was not reached for a while. There were escapades, in which she would find a willing co-conspirator and walk away from the convent towards a nearby commercial district. However, the escapades ended when neither she nor her companion could find their way back and they luckily were escorted home by police.

So my sister, the nun, faded into eternity. She had lived a life of accomplishments as a teacher, had completed a doctorate in philosophy, and seemed generally happy.

Yet her talent as a musician, her lovely voice and her social potential were traits which did not seem sufficiently fulfilled on this earth.

And she did live out her vow of celibacy, which meant no husband, no kids and no grandkids.

In the end, she seemed a bit lonely.

So what would she expect of the after-life? I never thought to ask her – probably because I fully assumed that she would give me the standard answer: an eternity of bliss in the company of God.

That stock answer does not seem to satisfy some of the other siblings.

Two Siblings Who Don't Fear Death

I don't wish to intrude into the privacy of two living siblings, who have lately expressed a certain nonchalance about dying. So I won't go into detail about their lives.

But I can say that in neither case does there seem to be any pathology whatsoever. They both seem contented, surrounded by children and grandchildren, and economically comfortable, though very modestly so.

Neither has expensive tastes. Neither seems to think he or she is in a position to change the world, during the time each has left on this earth. Neither seems worried particularly about dying nor particularly excited about eternal life.

Perhaps this revelation should not startle me. Perhaps I should have assumed that many of my own close family members are in that rather large category of good and decent folks who don't count much on eternal life as a final resting place. Or at least don't dwell on it as desperately as I do.

But it does startle me. Here's why.

Either we humans are wired to yearn for eternity or we are not. There is really no logical in-between condition, is there?

Again, I don't want to go on a tangent to explore the pathology of those who take their own lives. People who suffer from depression (particularly the acute episodes of bipolar depression) don't have the capacity to think rationally. They lose a job, or a loved one and find themselves in what Winston Churchill called the "black dog of depression."

I have never been depressed for more than a couple of minutes (typically after losing an election). I cannot grasp the abject depths of a feeling in which there is no light at the end of the tunnel, and in which the tunnel is a constant, painful, incomprehensible and unimaginably dark condition that grows more painful with each step.

I am not discussing suicide.

I am discussing a simple, apparently common insouciance about death, or what follows death, on the part of perfectly sane and perfectly well-adjusted adults.

I take note that many religions don't have a strong component of after-life theology. Many, if not most, Jews don't believe in an afterlife. Salvation consists of leading a virtuous personal life, contributing to your community, and leaving a legacy of good deeds that will make life better for those who follow you, including and especially your own offspring.

My knowledge of Buddhism and Hinduism is too superficial to characterize their views of the afterlife.

As to the world's largest single religion, Islam, we do have some vivid descriptions of the afterlife, including some involving a rather exaggerated number of young virgins placed at the disposal of the male martyr who dies for his faith. Certainly, one can conclude that Muslims conceive of sexual pleasures after an honorable life and death.

I can see that, to a certain point. I can see having a new, young, sexual mate who gives me pleasure for a month or so; and then another, and another, for 72 months (six years). But for eternity?

I can also appreciate the novelty of understanding nature, of discussing relativity with Einstein or the atom with Planck. I can look forward to a political give-and-take with Churchill, Kennedy, Bolivar and Marti. If invited, I could get excited about sharing the podium with the A-team: testing my wits with Aristotle, Augustine and Aquinas.

I would love to talk engineering with Da Vinci and political science with De Tocqueville. And I would love to have a few drinks with Churchill and hear a mass celebrated by John Paul II, with music by Bach and Mozart.

I would love to meet Joan of Arc and Martin Luther King.

But could I do that for eternity?

How far into eternity before I would simply get bored? Frankly, if I don't solve this puzzle, I feel that my entire *magnum opus* might fall apart, like happens to a deck of cards when one removes the middle one in the bottom row.

I have to dig into my being and explain.

Why I Yearn for Eternity

As I said at the beginning of this little book, I am the product of a very religious family. The idea of heaven has been ingrained in me since I was little.

What that does, psychologically speaking, is very interesting; it has been the subject of many studies. All such studies have concluded that a caring, hopeful brand of religious faith is a very positive factor in mental and physical health.

For the most part, that has been my experience. The idea that my good deeds in this life will be rewarded by an infinity of happiness has been a motivation beyond any other one that I can even begin to consider. It has carried me through tough times, when things did not go well. It is less of a motivating factor when things are going well, which is most of my entire life and all of my recent life.

At this point, as I approach my seventh decade of existence, other, less tangible things come into my psychological radar.

As with the Greeks, I think a lot about my legacy, or the reputation I will leave behind. As a politician, I think about that, though I am realistic enough to realize that most local politicians are forgotten the day after they retire or lose their last election. Perhaps you will get a park or a street named after you.

Big deal.

A more concrete legacy is one that comes from the Jewish tradition, which emphasizes the need for one's offspring to continue the good works of the deceased. I like that sort of legacy too, since it gives me a lot of satisfaction to think that my good works will be emulated by future generations. That idea motivates a lot of my writing. I would say it is the biggest motivator, since I am clearly not profiting or becoming famous from what I write.

But neither the Greeks nor the Jews with their concept of legacy suffice to make me whole with the concept of eternity. I need to believe something along the lines of what St. Anthony of Padua said: I need to believe that the after-life is compatible with life on this earth, only much, much better.

Part of the reason why I need to believe that is the many people around me who have gotten the proverbial "short end of the stick." The people who have been short-changed in their earthly existence.

If there is a God, he/she must have a special place reserved for the child who is blind, the soldier who comes back wounded or blind or paraplegic from fighting to protect our freedoms; the severely autistic child, the person who suffers from bipolar depression and ends up taking her life; people like Mother Teresa and Pope Francis who voluntarily give up the good things to minister to others; my niece who gave up her creature comforts to care for an unwed mother of two kids; the battered wife and the isolated, unaccepted, struggling refugee who leaves with meager possessions and has to scratch out a living in a strange land. If there is a God, he/she must have a reward for the Holocaust victim and for those imprisoned in the Gulag archipelago and those killed by starvation in the killing fields of Cambodia, Laos and China.

Otherwise God is not fair. And that is a logical impossibility.

Heaven is necessary to even the score for those of whom Emma Lazarus wrote her immortal poem:

> *Give me your tired, your poor,*
> *Your huddled masses yearning to breathe free,*
> *The wretched refuse of your teeming shore.*
> *Send these, the homeless, tempest-tost to me,*
> *I lift my lamp beside the golden door!"*

I think about the homeless a lot. How can people fall through the cracks of a "Great Society" like Miami's and end up in the streets, homeless and alone? How can there not be a good Samaritan for each one of those?

My heart anguishes for a while, and I discuss it with an older friend, a great economist who is also a follower of Jacques Maritain.

He is a Christian Democrat.

Oscar Arias, winner of the Nobel Peace Prize, is a Christian Democrat. Eduardo Frei of Chile was a Christian Democrat. Konrad Adenauer of Germany, who led that great nation back into the community of nations, was a Christian Democrat. Guy Perez-Cisneros of Cuba, who wrote (in collaboration with the great Haitian diplomat, Emile St. Lot), large portions of the United Nations Charter of Human Rights, was a Christian Democrat.

If you fuse the teachings of the Hebrew prophets and of Jesus onto the writings of the great political philosophers, the writings of John Stuart Mill and Alexis De Tocqueville and Jefferson and Hamilton and Madison and Jacques Maritain, you would get the modern equivalent of a Christian Democrat. It is, philosophically speaking, the combined work of people whom David Halberstam called (referring to the Kennedy administration) "the best and the brightest."

Without these people, and without the prophet who inspired them, we probably would not be a free nation, in a mostly free world.

That is my educated guess. But it doesn't solve the issue of what we can expect of an eternity in the company of an infinite, loving being. For those of us who walk the earth, in this era of relative prosperity and tranquility, will there be a change of circumstances that is worth even occasionally sacrificing oneself, turning the other cheek, walking the extra mile, when it is so much easier to feed one's narcissistic instinct?

That is the final question. And to find answers, we need to resort to the most enigmatic book in the Bible: Revelation.

Revelation

If you believe this otherwise inscrutable biblical source, the first thing that happens, after death, is that God evens the score. The book of Revelation is instructive here, in relation to those who have suffered much. All of their sufferings, says the prophet, will be recompensed by God:

> *He will wipe away every tear from their eyes, and death shall be no more, neither shall there be mourning, nor crying, nor pain anymore, for the former things have passed away.*

The score is also evened for the bad ones, and that includes all of us. Every time that we despised someone, or treated a weaker human harshly without rhyme or reason, or failed to come to the aid of someone being mistreated, or preferred our own wellbeing over the person who was hoping for the crumbs falling from our table of riches and glory, we created a hole in our personality that needs to be filled in with love.

O.K. I buy that. I can accept some painful purgatory or transitional period in which I will have to suffer some to acquire the kind of solidity that it takes to enjoy proximity with God. I can accept some form of Purgatory, though *I sense that for many people, their Purgatory is pretty much completed on this earth.*

My parents are both in that category – though for entirely different reasons. My father because of pride; my mother because of narcissism.

I share in both of those defects, and wonder if my Purgatory will be as painful as theirs, which I saw firsthand.

But this chapter is about heaven, not hell. And I still don't have a rational, understandable, believable theory of eternity. I am not sure I ever will; and here's why.

The Unthinkable Importance of Eternal Life

In the final analysis, mankind is destined to having to rely on faith – as far as an after-life. I don't mean faith in the Jesus story, which is quite believable and even scientifically consistent with our best calculations of how humans should behave. By all accounts, using all our best equations and quantitative models, a Judeo-Christian life is the best prescription for individual and collective happiness on this earth.

But here I am dealing with something totally different, though related. Here I am dealing with a mystery that lies outside of science. Eternity is an important reality, but it is not susceptible to scientific analysis, yet it lies in the periphery of all scientific equations. Eternity is most likely our fate, and yet we are doomed to never fully grasping it.

But my instinct drives me a little farther. Deep inside my consciousness, in the part of me that is willing to consider death, I yearn for eternity. I don't want to die.

My guess is that my dilemma is easily solved. The reason I cannot understand eternity is that I am not anywhere close to it. Right now, I am a very limited cog in a very big wheel; I am one of thirteen legislators in a commission that approves, but does not formulate, a seven-billion-dollar budget. I yearn for executive power, so that I can show how the resources of this city-state (Miami-Dade) can make life tolerable for all and great for the bulk of our citizens.

As the writer of five books, and countless articles, I feel my message has not reached a sufficient audience. My 10,000 social media followers are not enough to convince me that my ideas have taken root. I want to live at least another quarter century.

As an earthbound human, blessed with talents and opportunities seldom offered to one person, I sense that my mission is far from being accomplished. My life expectancy, at my present age of 68, is less than a decade.

But I don't look at it that way. Not at all.

In my estimation, my adult life is barely halfway completed. I hold to a theory that says my adult life is only half through. For the last couple of decades, I calculated my life expectancy based not on averages, but on maximum life spans. For men, that is 114 years.

I have lived a little less than half a century as an adult. I figure I have another half century left to live as an adult.

But maybe not. Maybe I am average and not optimal. Maybe I will die in the next ten years, as predicted by probability analysis. And so I face three possible scenarios: another half century (very unlikely), another decade (statistically sound) or another few days, if something catastrophic were to happen to me.

Science and the Theory of God

All of those are worth considering. Correct that: all of those need to be considered for any kind of scientific analysis.

The scientist in me wants to harmonize these scenarios and make sense of all of them. Something powerful inside me tells me that all three scenarios are wanting. All three are sorely lacking, compared to what I yearn for.

I want to take my best shot at eternity. But to understand eternity, I have to harmonize my destiny with that of my youngest brother. For harmony, for logic, for completeness of my theory of life and death, I have to discuss my youngest brother.

The Last Days of My Youngest Brother

As I was writing these lines, I was reminded by my brother Mel of the last days of another, our youngest brother. His name was Fred.

I don't wish to over-dramatize the life of my youngest brother. I do want to learn from it. Fred was a victim of cancer; he was the youngest of fourteen and had the shortest life. He died at 57.

As with my other siblings, I want to spare the private details of a difficult life. Like my oldest sister, who was celibate, Fred never bore biological children. He married a nurse who was a single mom, and whose son Fred adopted and loved.

I was reminded by another brother that Fred changed a lot of lives, particularly in his last days. Every sibling in a family of close to fourteen (minus the nun who had died and a sister who was herself very sick) made an effort to visit Fred as he battled esophageal cancer. Analyzing it now, I realize that Fred needed that.

And so did we.

Earthly life, for most of us, is a maddening combination of relationships. We give and we take; we comfort and we yearn for a comforting hand; we see peers who succeed where pure merit would not support such triumph; we see others who suffer illness and deprivation that is totally undeserved.

Eventually, we come to the realization that our share of earthly happiness is darn good, all things considered. But we wonder if the formula works for the entire species – for all of us. For Fred, it doesn't seem fair that his life would end as a body ravaged by cancer and devoid of all hope.

Fred deserves an afterlife.

I am not so sure I deserve it. But I sure yearn for it, among other things because I want to see Fred happy and fulfilled. It is Fred, more than any other single human I ever met, who evokes in my mind the quote that begins this chapter, by Pierre Teilhard de Chardin. I want to recreate myself in the company of Fred, remember and replay the good times – in particular the days during Christmas when we would hear the carol called "The Drummer Boy."

I want to see every child that was born without sight, or missing limbs or not fully intellectually developed reach, before my very eyes, the full flowering of mind and body.

The End of Times

As I pondered the final observations of this little book, crafted as it is from insights of better writers and wiser thinkers, I was given an article to read. It was written more than a decade ago, and it compared the famous Civil War battle of Gettysburg with the tragedy known as 9/11, and in particular the hijacked airplane that crashed in a plain in Pennsylvania. (The article was by famed humorist, Dave Barry.)

As we have come to know, a handful of passengers on that doomed flight were able to speak or leave text messages to loved ones on the ground. The messages give us a glimpse of good, decent people, determined to die with the kind of special dignity that belongs to those who resist evil, and fight for their fellow human beings, even if they are total strangers.

In heaven, if there is a heaven, I would love to chat with those heroes. I would want to pick their brains, scour their souls, share their despair and revel in their heroic resolve. They are not my relatives or my friends, but I would like to befriend them and treat them like brothers and sisters.

Could this kind of new friendship motivate me for all eternity? Or would I soon become satiated, and even bored, in the company of fellow creatures?

Eternity is a long time, when compared to my 68 years. How interesting can it possibly be, after the first 68 years in the company of loved ones and of the creator?

How interesting can it be, after the first 680 years, or after the first 680,000 years?

It is a quandary that might not be solved, or solvable, for my limited mind. It might be as inscrutable as the Big Bang itself, the inflationary moment immediately after it, and the theoretical eternity that extends back before the Big Bang and that presumably explains it.

I might never know.

But I sure would like to find out if Fred ever got his fair share of glory. And I sure would like to meet those heroes in that plane that crashed in Pennsylvania, close to Gettysburg.

Quantifying Eternity

There is no way to quantify eternity. But I will try. It is my nature to quantify. I am an engineer, after all.

I have come up with two equations that might fit the eternal parameters. One describes the phenomenon that occurs as stars get drawn into a black hole. Without getting too technical, there is now some agreement that inside most galaxies, there is a "super-heavy" black hole, which is nothing more and nothing less than a bunch of matter, packed in like sardines, into a very small space. A black hole attracts light with so much force that it cannot escape.

Anyhow, the point is that in many galaxies there is this immensely powerful magnet that makes stars close to it spin faster and faster, until they reach unthinkable speeds, at which their mass increases to unthinkable size.

Heaven is probably like that. As we get closer to the greatness of God, we are propelled to greater and greater dimensions of joy, of understanding, of appreciation and love for God and his/her creatures. We can never reach the super-heavy black hole itself, because the energy required to get us there is increasing exponentially, along with our mass and inversely to our size. We get more massive, more intelligent, more loving, and slower in our journey, as we get closer to God.

Another equation might help my amateur reader. It is one that describes a mathematical curve, that defines an object which has infinite sides but a finite volume. It is called the Torricelli curve and also the horn of Daniel, because it is shaped like the proverbial horn of plenty. If a bucket could be made with sides that are shaped like Torricelli curves, its volume would be finite but its sides would be impossible to paint, due to their infinite area.

Like black holes and Torricelli curves, the equations that define God are impossible for our minds to grasp. Our minds struggle with the physical reality described by the equation.

The quantitative interface of God with man is like those mathematical equations, which describe physical realities. We are finite. God is infinite. For whatever reason, God invites us to spend eternity with him/her. How exactly it all works out will remain a mystery.

But if it's true, there is no doubt that it will be fun to explore it.

Until the end of time.

AUTHOR BIOGRAPHY

Xavier L. Suarez is an author, lawyer, and politician. He served as mayor of Miami in from 1985 to 1993 and currently serves as a Miami-Dade county commissioner. His education includes degrees in mechanical engineering from Villanova University, a juris doctor from Harvard Law School, and a master's in public policy from Harvard's John F. Kennedy School of Government. He is married to Rita Suarez and has four children.

www.ingramcontent.com/pod-product-compliance
Lightning Source LLC
Chambersburg PA
CBHW030929180526
45163CB00002B/511